Lecture Notes in Economics and Mathematical Systems   620

Christian Artmann

The Value of Information Updating in New Product Development

 Springer

Dr. Christian Artmann
Immenstadt, Germany
christian.artmann@whu.edu

ISBN 978-3-540-93832-3   e-ISBN 978-3-540-93833-0

DOI 1 0.1007/ 978 -3-540-93833-0

Lecture Notes in Economics and Mathematical Systems ISSN 0075-8442

Library of Congress Control Number: 2008943983

Cover design: SPi Publishing Services

Printed on acid-free paper

springer.com

To my family

Preface

Ever since I learned to fly airplanes, I am passionate about long-distance gliding in the Alps. Every time I take off in the early morning, I never know exactly which challenges I will have to overcome during the day. Especially the weather conditions are constantly changing, making it impossible to forecast even the thermal behind the next mountain range with certainty. Despite this lack of information, I have to decide whether it is worthwhile to "jump" over the ridge into the next valley towards my target. To obtain the experience how to instantly study the new conditions, reassess the previous plan, and adjust the route accordingly was a tough process, ending several times in an acre instead of my home airport. But most other times, when it worked out, I was rewarded with great moments and majestic impressions.

New product development projects resemble long-distance flights in many aspects. They are also exposed to numerous uncertainties and require continuous adjustments of the development efforts in order to successfully launch the product. While decision making during a flight is relatively straightforward and can be based on experience, it is far more complex for development projects. The basic principle, however, remains the same. Additional information obtained during the process has to be considered in the decisions for the remaining periods. Although this process is intuitive and follows the natural way of making decisions, it has hardly been formalized for development projects so far. It became the goal of this work to develop a decision model that explicitly takes information updates into account in order to better manage uncertainty. I hope it will be instrumental to managers and academics alike when facing difficult decisions in complex new product development projects. This work was submitted in fulfillment of the requirements for the doctoral degree at WHU, Otto-Beisheim School of Management, in the department of Production Management, chaired by Professor Dr. Arnd Huchzermeier.

I would like to thank Arnd Huchzermeier for his guidance and the inspiring environment he creates at his department, including the opportunity for me to study and research at Stanford University. I am deeply indebted to Rainer Brosch for the great intellectual and personal support during all these years, and for the many unforgettable moments. Furthermore, my thank goes to Daniela Schmitz-Wiehenbrauk and Christoph van Wickeren, who have been great colleagues and friends, making my time enjoyable not only at WHU. I also thank Stefan Spinler who was ready to help whenever necessary. I owe special gratitude to Georg Strasser who has been a dear friend since studying together in Karlsruhe. Driven by his expertise and enthusiasm, he added different perspectives and ideas in the many enriching discussions I had with him. Moreover, I appreciate the financial support from the German Ministry of Education and Research that funded this dissertation.

Finally, I am most grateful to my parents, Helmfried and Marlene, who have always encouraged and supported me throughout my life. And to my brothers and sister, Volker, Petra, and Marko, who are always with me. You all have consistently helped me keep perspective on what is important in life and added a lot of joy. This work is dedicated to you.

Immenstadt, August 2008   *Christian Artmann*

Table of Contents

List of Figures

List of Tables

List of Abbreviations, Variables, and Functions

*	(indicates optimality)	G	Gamma distribution
α_t	Improvement cost at stage t	I	Initial investment cost
		IG	Inverse Gamma distribution
a_t	Managerial action (i.e., continue, improve, or abandon) chosen at stage t	i.e.	id est (that is to say)
		$k(a_t)$	Expected performance state improvement if action a is selected at stage t
c_t	Continuation cost at stage t		
cf.	confer (compare)	μ	Mean of market performance requirement distribution
d	Market performance requirement		
$E(\cdot)$	Expectation operator	M	Upper bound of profit margin
e.g.	exempli gratia (for example)	m	Lower bound of profit margin
Eq.	Equation		
et al.	et alii (and others)	N	Normal distribution
$f(\cdot)$	Distribution density function	N_k	Multivariate normal distribution (k dimensions)
f., ff.	following	NPD	New product development
γ_t	Cost of market performance requirement update at stage t	NPV	Net present value
		OV	Option value

$\Pi(x)$ Expected market payoff if launched product has performance level x

$\Pi(x, z)$ Expected market payoff after update with signal z if launched product has performance level x

P Market payoff

$P_t(x, z)$ Posterior project value in state x at stage t if prior managerial policy is applied

$P_t^\tau(x)$ Expected project value in state x at stage t for an update at τ if prior managerial policy is applied

p., pp.   page(s)

$\Phi(\cdot)$ Cumulative probability function of a distribution

post   (superscript) Posterior value, i.e., after update with signal z

prior   (superscript) Prior value, i.e., before update with signal z

QFD   Quality Function Deployment

QR   Quick Response

r Discount rate

R&D   Research and Development

Σ Covariance matrix

σ Standard deviation of market performance requirement distribution

$\check\zeta$ Prior standard deviation of market requirement mean

St Student's t distribution

St_k Multivariate t distribution (k dimensions)

τ Updating point in time

θ Unknown parameter of market performance requirement distribution (true market performance requirement)

T Project duration

$tr(\cdot)$ Trace of a matrix argument

$V_t(x)$ Project value in state x at stage t (no updating possibility)

$V_t(x, z)$ Project value in state x at stage t after update with signal z (posterior project value)

$V_t^\tau(x)$ Project value in state x at stage t in expectation of information update at time τ ($t < \tau$) (expected project value)

$V_t^I(x, z)$ Value of information update with signal z in state x at stage t

$V_t^{I,\tau}(x)$ Expected value of information in state x at stage t for update at time τ ($t < \tau$)

Wi_k Wishart distribution (k dimensions)

X_t Performance state of project at beginning of stage t

ω_t Development uncertainty at stage t

z Observed market performance requirements in follow-up study (signal)

1

Introduction

1.1 Motivation and Problem Definition

At the beginning of the 21st century, the effects of the proceeding globalization are clearly visible. Almost every company in one of the old, established economies is affected by it and faces increased competition. While these firms competed in fairly stable markets against known rivals for a long time, the pattern has significantly changed over the last decade. Companies in emerging markets are increasingly becoming capable of producing goods at comparable quality, but at significantly lower costs due to advantages in wages, taxes, or (environmental) regulations compared to their established competitors. Responding to these developments by outsourcing business processes and shifting less knowledge intensive activities like production or assembly to low-wage countries has only a limited, short-term effect in a globalized world. Companies are therefore increasingly recognizing that their competitive advantage rests on the capability of continuously developing innovative products and launching them successfully in the market (Holman et al. 2003).

Best performing companies are able to generate up to 50 percent of their sales and profits from new products (Cooper et al. 2004). However, even among this group of firms, most of this share results from minor new product development projects (NPD) such as line extensions, improvement of existing products, or cost reductions. Only 10 percent of their NPD projects belong to "new-to-the-world" true innovations although most of the firms' research and development (R&D) resources are generally dedicated to these projects (cf. e.g., Adams and Boike 2004; Cooper et al. 2004). The key reason for this low cost-benefit-ratio is the extremely high failure rate of such projects. Studies report an average success rate for innovation projects of only about two to five percent depending on the particular industry (Nussbaum et al. 2005). The average for NPD projects is not significantly higher

C. Artmann, *The Value of Information Updating in New Product,*
DOI: 10.1007/978-3-540-93833-0_1, © Springer-Verlag Berlin Heidelberg 2009

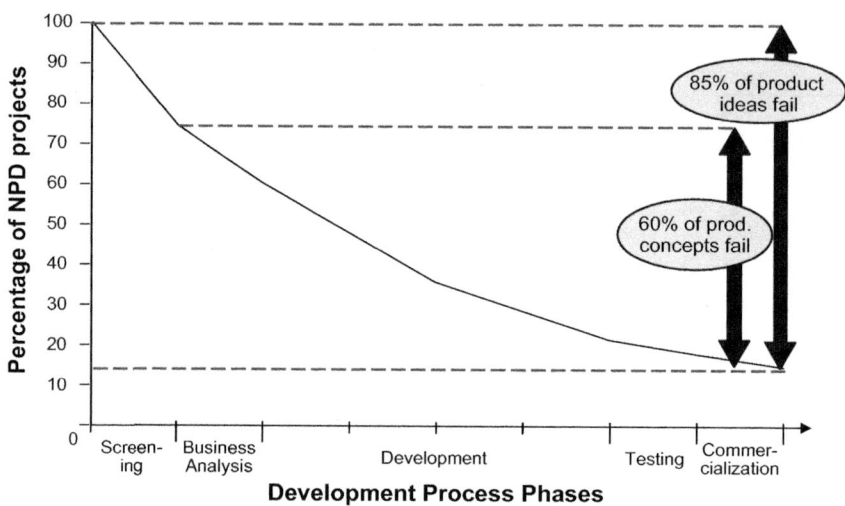

Source: Griffin (1997a).

Fig. 1.1. Failure rate of NPD projects

(Griffin 1997b). Almost 25 percent of all new product ideas fail during the initial screening phase of the NPD process before a detailed business plan has been developed. Another 60 percent of the further pursued concepts fail during the actual development process when already significant expenses have accrued (see Fig. 1.1).

The reasons for the high failure rates of NPD projects are manifold. Besides organizational, process, engineering, or financial issues, it is particularly the underlying uncertainty of NPD projects which leads to unforeseeable obstacles during the development process and hence, to high failures (Sommer and Loch 2004). The uncertainty can hereby result from various sources, e.g., market payoffs, project budgets, product performance, project schedule or market requirements (Huchzermeier and Loch 2001). The latter uncertainty has been identified as highly critical (cf. e.g., Ottum and Moore 1997; Mishra et al. 1996). It can be traced back to insufficient customer orientation and market research not only during the early phase, but also throughout the entire development process of a NPD project. This is especially important for companies in fast-paced industries (e.g., consumer electronics or high-tech) where it is almost impossible to forecast the requirements of potential customers precisely at the beginning of a project when the development goals have to be set. If, however, the market requirements are not met with the launched product or a competitor introduces a comparable product at the same time, the potential revenues from the development project shrink dramatically and hence, the entire profitability of the project is threatened. The timely identification and evaluation of such market trends in order to update and revise initial goals during the development process is therefore the key to reduce (market) uncertainty in NPD projects (Atuahene-Gima 1995).

Although information updating is a well-known method for uncertainty reduction in the field of decision theory (cf. e.g., Berger 1985), it has hardly been applied in the area of NPD. Much of the steadily increasing research in product development and innovation management has been conducted in fields like organizational science, marketing, industrial design, engineering, accounting, or strategy (Ulrich 2001). The focus in these areas has primarily been on organizational and cultural success factors, on the development process and its improvement, the study of strategic issues like project or technology selection, or the identification of customer needs. Relatively little research attention has yet been paid to quantitative models that address the characteristics of NPD projects, like the effect of different sources of uncertainty on the value of managerial flexibility or the reduction of uncertainty through information updates. Although empirical research stresses the need for more formal and quantitative models to better manage the involved uncertainty in the development process of new products (cf. e.g., Mahajan and Wind 1992), the development of such models has not been a major theme in disciplines like management science or operations management (Wind and Mahajan 1997; Krishnan and Loch 2005). Only recently, the number of quantitative models for decision making in the area of product development seems to increase (Krishnan and Ulrich 2001).

One reason for this trend is the growing recognition that some of the NPD problems are similar to those already addressed in other areas of operations management, like supply chain or risk management for example (Seshadri and Subrahmanyam 2005; Loch and Terwiesch 2005). Hence, some of the developed concepts and methods for these problems, like the information updating approach, might also be applicable to product development projects. For example, the above described problem of making development decisions in the presence of high uncertainty about the customers' performance requirements has analogies to supply chain management decisions under demand uncertainty (e.g., Fisher and Raman 1996; Eppen and Iyer 1997). There, management faces the following challenge: In industries with long lead times, like the fashion industry for example, retailers have to place their orders far ahead of the selling season. Thus, at the ordering point in time, they experience great uncertainty about the actual demand. Ordering too little results in unsatisfied demand and hence, lost profit, while too much inventory at the end of the season requires costly markdowns. An optimal response to the uncertain demand based on traditional forecasting methods is hard to achieve. To improve the situation for the involved mem-

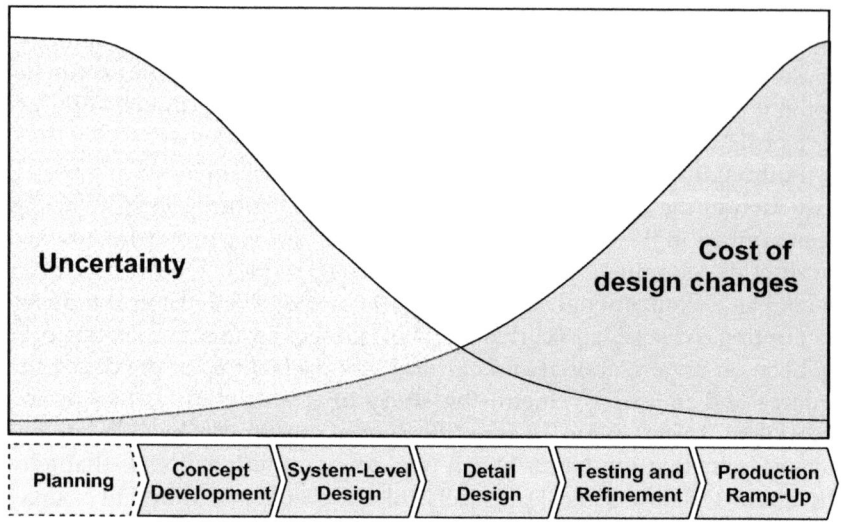

Fig. 1.2. Managerial trade-off: Uncertainty reduction vs. cost of design changes

bers in the supply chain, so-called Quick Response (QR) models have been developed which allow to postpone some orders until more demand information becomes available. More specifically, the retailer places a first order that is produced at relatively low costs far ahead of the selling season based on best estimates. Later on, when information regarding the actual demand can be gathered from related items or from orders placed at fashion shows, this information can be used to update the initial demand estimates and adjust the more expensive second order quantity.

This decision problem closely resembles the ones occurring in NPD projects where the company has prior to the start of the project high uncertainty about the actual market requirements. Management has therefore to base its decisions about the product specifications or the appropriate design on available market studies, forecasts, or best estimates. It is in the nature of such information that it is not very precise at this point in time. But as the development project progresses, additional information becomes available or can be generated (by additional market studies for example) to update the prior beliefs and to adjust or revise earlier decisions. This information gain, however, is opposed to the cost of design changes that generally increase significantly over time (e.g., Zangwill 1992). Thus, as Fig. 1.2 depicts, the trade-off is – similar to the one in the QR models – to make either an early decision at relatively low cost but with little information or to wait until more information becomes available at the expense of higher cost (Loch and Terwiesch 2005). This decision problem comprises questions like the optimal timing of an information update, its value, and the optimal managerial response to the obtained information gain.

1.2 Research Objective and Modeling Approach

This thesis addresses the described problem of decision making under uncertainty in NPD projects by developing a general valuation model that takes the managerial flexibility into account to respond to two sources of uncertainty: Firstly, technical uncertainty stemming from the performance variability of the project. It represents the uncertainty whether the initially targeted product performance can actually be met. To respond to these unforeseen contingencies, we assume that management has the choice between three possible actions at each review point of the project: Besides simply continuing the project, the company can either improve the performance by investing additional resources or abandon it if any further investments are not justifiable. Secondly, we further explicitly consider market uncertainty in form of market performance requirement variability in the model. Management has the possibility to reduce this type of uncertainty by acquiring additional information during the development process and hence, to update its initial forecasts. The optimal managerial policy can then be adjusted to the updated estimates of the market performance requirements.

The possibility to reduce uncertainty and respond accordingly by adjusting the optimal managerial actions has substantial value. Real options theory provides a rigorous and well-known framework to evaluate such projects with operational flexibility (cf. e.g., Dixit and Pindyck 1994; Trigeorgis 1996). However, most real options models only value the existing options of a project at different review points. They do not explicitly take the value of additional information into account since they regard the incorporation of information as a consequence of the project progression (Miller and Park 2005) and not as an active possibility to enhance the decision basis and thus, the project value. To also consider the latter possibility of a managerial response, we derive a general Bayesian updating formulation for the above described market requirement information that takes the characteristics of NPD projects into account and integrate this mechanism into a real options valuation framework.

With this approach, we provide – to the best of our knowledge – the first decision framework that combines statistical decision theory in form of Bayesian analysis with a real options framework for NPD projects. The developed model allows to determine the value of projects that are exposed to the above mentioned sources of uncertainty. In addition, the value of an information update, the (in expectation) optimal updating point in time as well as the optimal managerial decisions in response to this uncertainty reduction can be determined. This analysis allows to explore the conditions under which an update is beneficial and to study the effects of the information update on the optimal managerial policy over the entire development process. Simultaneously, the benefit of higher initial investments in design flexibility, which reduce the later costs of corrective actions, can be analyzed and a threshold value for these investments derived. With this framework and the derived insights from its analysis, the thesis contributes to close the above mentioned gap of information updating methods in NPD decision models as well as the general lack of quantitative methods in this area. It thus takes a step forward enhancing the valuation and the management of the increasing uncertainty in product development projects.

1.3 Structure

The remainder of the thesis is structured as follows: In Chapter 2, we review the relevant literature which includes new product development, real options analysis, and decision theory. We summarize the key contributions, discuss their implications for our research objective, and identify aspects in which our approach differs from the existing ones. Chapter 3 contains the formulation of our valuation model. We start with the presentation of a real options model which, with some modifications and extensions, builds the basis for our framework. Then, the second cornerstone in the development of our model, the information updating mechanism, is introduced by deriving a general Bayesian updating formulation for our problem setting. We explicitly model the update of the mean, the variance, and both, the mean and the variance of the normally distributed market performance requirement distribution. In the last section of this chapter, we combine these two concepts and develop the overall valuation model. At each of these steps, we also derive, in addition to the one-dimensional performance parameter case, the multidimensional model formulation in order to ensure the practical applicability of our framework.

In Chapter 4, we analyze our model and derive properties in closed form, which include the impact of an information update on the project value and the conditions under which it will be most beneficial to obtain additional information for such an uncertainty reduction. In Chapter 5, we resort for those characteristics which elude an analysis in closed form to a numerical study that is based on a real-life NPD investment project. This analysis in a comparative static manner provides insights on the change of the optimal managerial policy compared to the basic decision model if a later updating possibility is considered. In addition, it illustrates the impact of cost structure changes to facilitate late development changes and, by relying on real data, simultaneously demonstrates the practical applicability of our model.

Finally, Chapter 6 summarizes the main findings, discusses key contributions to the current research in this field, and provides clear managerial insights. In this conclusion of our study, we further discuss possible extensions of the developed model and identify directions for future research.

2

Literature Review

As indicated in the preceding chapter, our research builds on concepts that have been developed in previously unlinked areas. We therefore have to draw on different strands of literature that can be subsumed under the following categories: new product development, decision models with information updating, and real options theory.

This chapter surveys these literature strands and reviews the relevant concepts and theories. We start with the literature on new product development, providing an overview of key development project characteristics as well as the success factors identified by empirical studies. Thereafter, decision models with information updating possibilities are reviewed. Besides giving an overview of the relevant information updating modeling approaches, we study corresponding decision frameworks developed in different areas of operations management and discuss their applicability to our research objective. Finally, the key concepts of real options theory are presented, including different valuation methods and insightful frameworks for research and development applications.

2.1 New Product Development

This section presents the key characteristics of new product development projects and reviews the (empirical) literature with respect to identified challenges as well as success factors of managing the project inherent uncertainty. The literature on this topic is vast since it encompasses marketing, engineering, strategic, as well as organizational and behavioral aspects. Given our research objective, we will thus primarily focus on uncertainty reduction and information generation related approaches from an operations

C. Artmann, *The Value of Information Updating in New Product,*
DOI: 10.1007/978-3-540-93833-0_2, © Springer-Verlag Berlin Heidelberg 2009

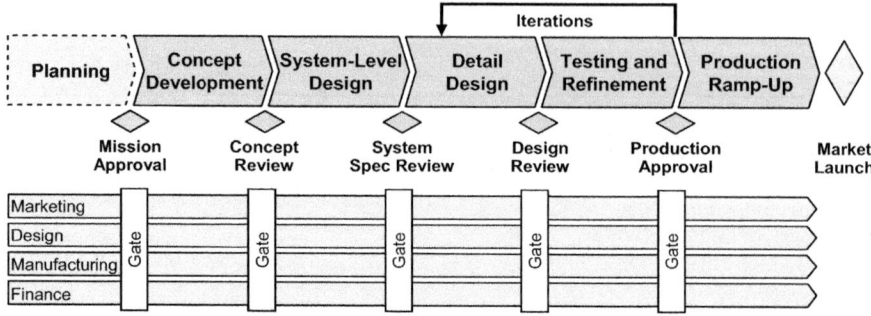

Source: Adapted from Ulrich and Eppinger (2004).

Fig. 2.1. Stage-gate development process

management perspective.[1] The obtained insights will highlight the need for quantitative decision models in order to enhance the understanding and the management of uncertainty in the area of NPD. In addition, they will illustrate the basic characteristics of NPD projects underlying our framework.

2.1.1 NPD Process Characteristics

The development of a new product generally comprises all activities starting with the identification of a market opportunity and ending with the launch of the product into the market. It is therefore more specific and targeted than general research activities (cf. e.g., Brockhoff 1997a, p. 35 ff.). The generic process of NPD is a sequential approach driven by the chronological progression of development tasks. The first formal schemes, today often referred to as the phased review process, elaborate on the physical sequence and brake up the development activities in discrete phases, each ending with a review point where decisions about the further progression and its funding are made (Cooper 1994). Thus, this formal process has almost solely a project management focus on the engineering activities in order to ensure the completion of the project on time, within specification and budget. It further addresses only technical risks and ignores relevant tasks in marketing or finance as well as any interactions between these different functional disciplines.

These deficiencies have been addressed in the so-called stage-gate process models (Fig. 2.1). Resembling somewhat the phased review process,

---

[1]For a more general overview of the product development literature, the interested reader is referred to excellent reviews of Krishnan and Ulrich (2001) (decision making), Griffin and Hauser (1996) (marketing aspects), Brown and Eisenhardt (1995) (organizational perspective), Montoya-Weiss and Calantone (1994) (environmental and contextual variables), or Ernst (2002) (general success factors).

they also break the NPD process up into a predetermined set of stages with predefined checkpoints (gates). The difference to the previous models is that each stage comprises clearly prescribed cross-functional and parallel activities. The gates after each stage contain deliverables for each functional area that the project must pass in order to proceed to the next stage. Typically, between four to six stages are found in industry (Cooper 1994). The first stage, planning, builds upon advanced research and development activities by investigating the market potential of the product concept, exploring possible product architectures and manufacturing methods, as well as conducting financial studies. The results of these analyses build the basis for the business case of the project which sets the aimed project specifications, defines the required tasks, and describes the budget and schedule constraints. After the formal approval of the project, these aspects are refined and further explored during the concept development stage before the actual design activities are started (system-level design phase). Over several building, testing, and reworking iteration cycles, the product advances, reaches manufacturability, and finally, during the production ramp-up stage, the readiness to be launched into the market.

By integrating all involved departments into the development process, the stage-gate model is highly cross-functional, has a strong market orientation, and fosters a holistic assessment of the NPD project over the entire process. Besides technical aspects, management is thus urged to also assess market, financial, and legal aspects of the projects on an ongoing basis. In addition, with precisely defined deliverables and clear go/no-go decision criteria at each gate, this approach encourages task completion and decision making. It thus builds the basis to deal effectively with market and technology uncertainty surrounding NPD projects (Lint and Pennings 2001; Griffin and Hauser 1996).

To further increase the efficiency of the stage-gate process, several refinements of this approach have been proposed focusing on speed, flexibility, and more efficient allocation of development resources. Takeuchi and Nonaka (1986), for example, stress the need for addressing the continuous interactions between the members of the multidisciplinary team and the parallel processing of tasks in the development process. Their process has therefore overlapping stages where operational decisions are incrementally made within the teams. Strategic decisions of the project, however, are delayed for a more flexible response to market changes. Similarly, Cooper (1994) proposes fluid and overlapping stages to shorten cycle and development times and to account for parallel development efforts like concurrent engineering. In this process, the gates are not fixed anymore, but fuzzy in order to allow for conditional and situational decisions. This avoids, for example, project delays when certain (minor) criteria in a functional area are not met. In other words, the so-called third-generation processes propose higher flexibility to address the project specific development characteristics and to account for the inherent uncertainties. Most companies have integrated such flexibility in different degrees into their conventional stage-gate processes.

But regardless of the degree of flexibility, parallelism, or fuzziness of the activities and gates, the in principle sequential nature of the major tasks and building blocks remains and requires still – if not even more – precise valuation criteria for go/no-go decisions. The development process thus evolved from a pure project management driven description of design activities to an information processing (Levitt et al. 1999; Loch and Terwiesch 1998) and risk managing system (Riek 2001; Büyüközkan and Feyzioglu 2004). From the latter perspective, the process starts with the identification of various risk factors, their evaluation, and prioritization. As the project progresses, however, these risks are generally reduced as uncertainty gradually resolves with technical problems being solved and more information about the market, e.g., customer requirements, sales volume, competitive environment, etc., is becoming available (Ulrich and Eppinger 2004). Thus, information generation and processing is highly linked with managing the project inherent risk.

2.1.2 Types of Uncertainty and Possible Responses

Development projects are exposed to numerous uncertainties. They can generally be traced back to the following sources: market, technical, resource, and schedule uncertainty (cf. e.g., Souder and Moenaert 1992; Huchzermeier and Loch 2001; Ulrich and Eppinger 2004, p. 20 f.). Market and technical uncertainty are often regarded as the most decisive ones while budget and/or schedule overruns are either induced by the former two sources of uncertainty or arise from managerial or organizational deficiencies. In the following, we therefore will solely concentrate on the former. Market uncertainty comprises, for example, customer requirements, moves of competitors, market size, or pricing. Technical uncertainty, on the other hand, relates to aspects like technology selection, design and product architecture, or the definition of product specifications. The overall degree of uncertainty is of course highly project specific and depends on aspects like the degree of innovation, selected technology, project duration, or the characteristics of the target market. Independently thereof, the overall uncertainty is generally highest in the early stages of the development process, often called the fuzzy front end, when the customer requirements and other market characteristics are still too ambiguous and many technical details have not yet been resolved (Dahan and Mendelson 2001; Schröder and Jetter 2003). In presence of such uncertainty, management has to adjust its development efforts accordingly.

Numerous approaches have been developed to respond to these uncertainties with appropriate concepts and frameworks. From a project management perspective, a well-defined and consequently implemented development process facilitates, as indicated above, the management of uncertainty. The precisely described tasks and valuation criteria at the different decision gates ensure that the project only progresses if all necessary issues have been sufficiently addressed. Several empirical studies stress that an implemented, formal development process is a key success factor for NPD projects exposed to (high) uncertainty (e.g., Griffin 1997b; Cooper and Kleinschmidt 1995, 1996). With respect to the uncertainty the project is exposed to, it is essential to adapt the generic development process accordingly to the firm's and the project's unique context and to choose an appropriate development strategy (cf. Ulrich and Eppinger 2004, p. 18. ff.).

Technical uncertainty can be reduced, for example, by exploring multiple solution paths in parallel. It is an expensive, but effective approach to ensure that one of the developed solutions succeeds and is therefore well suited for projects exposed to high technical uncertainty as Srinivasan et al. (1997) show. They provide empirical evidence that parallel prototyping resolves significant uncertainty in the mid to late stages of the NPD process. Dahan and Mendelson (2001) indicate that parallel development is also valuable in the early phases, i.e., to pursue multiple concepts in parallel and select the best design at a later stage. Postponing the finalization of the project specifications may be particularly beneficial in dynamic environments as Bhattacharya et al. (1998) stress. Where such an approach is either not applicable or too costly, an iterative development process with rapid design-build-test cycles may allow to resolve technical uncertainties in a fast and efficient manner (Thomke 1998; Smith and Eppinger 1997). Besides adapting the development approach to the technical problem setting, the project itself can be adjusted and setup in such a way that the inherent technical uncertainty is already a priori reduced. Decreasing the complexity by reducing the number of new parts or the diversity of applied core technologies lowers the technical development uncertainty and hence, increases the project success rate as Murmann (1994) as well as Meyer and Utterback (1995) empirically demonstrate.

Market uncertainty, on the other hand, can on a conceptual basis be reduced by shortening development lead time (Shelley and Wheeler 1991). A prominent approach therefore is the already mentioned parallelization of development tasks through implementation of concurrent or simultaneous engineering (Krishnan et al. 1997). With the reduced time span between the start of the development activities and the market launch when most of the uncertainty is resolved, a shorter time horizon has to be overlooked and hence, certain trends may already be observable at this project stages. A high development flexibility to address late market requirement changes and to postpone the final specification of the product is another approach to respond to uncertainty stemming from the market side. This can be achieved, for example, through a modular product architecture (Ulrich 1995; Krishnan and Bhattacharya 2002).

In case of extremely high uncertainty in NPD projects, so-called unforeseeable uncertainty, which prevents to recognize the relevant influence variables and hence, to plan ahead of time, the only two fundamental strategies are trial and error learning and selectionism (Pich et al. 2002). The former strategy tries to flexibly adjust project activities to new information as it becomes available at the costs of failures and project delays. Selectionism, on the other hand, involves pursuing several approaches in parallel and independently of one another and selecting the best one ex post. The costs of this strategy are also very high due to parallel activities (including bound resources) as well as forgone profits due to elimination of product variants. Sommer and Loch (2004) show that in presence of unforeseeable uncertainty and poor testing possibilities, trial and error learning should be preferred over selectionism. In case of perfect testing opportunities, both strategies offer equal results.

The extremely high costs of both approaches limit their practical applicability on NPD cases where uncertainty is completely unforeseeable. As stated before, such an uncertainty holds only true for a diminishing number of NPD projects. In most other cases, however, the influenceable variables and their functional relationship are known. Thus, less expensive approaches can be applied to reduce uncertainty. For these projects, which are also in the focus of our model, the timely generation and integration of information during the development process remains key to optimal decision making.

2.1.3 Information Generation and Updating

Numerous empirical studies stress the importance of timely identifying and evaluating external trends in order to update and revise current project targets during the development process. Atuahene-Gima (1995) and Mishra et al. (1996), for example, show that the generation of market information about current and future customer needs, competitive dynamics, and technology changes throughout the development process (and product life cycle) has a high impact on the future project success. These insights are supported by Balbontin et al. (1999), who study new product development success factors in American and British firms. In addition to the former two studies, they observe a high correlation between a company's forecasting activities and abilities (e.g., of the market potential and volume) and the NPD success. Finally, the findings of Cooper and Kleinschmidt (1994) stress the importance of observing the competitive environment in order to reduce this kind of market uncertainty and hence, to timely respond to competitors' moves by adjusting project targets. On the other hand, there are several large scale numerical studies who identified inadequate market analysis as well as lack of appropriate market data and updates among the top three factors for NPD project failures (cf. e.g., Cooper and Kleinschmidt 1996; Little 2004).

On the technical side, the feasibility of the selected product concept and architecture (Carbonell-Foulquié et al. 2004; Polk et al. 1996) as well as information about the development of the underlying technology (Iansiti 1995), especially for novel products, are regarded key to resolve technical uncertainty. Moreover, Loch and Terwiesch (1998) analytically show that this type of uncertainty, causing costly engineering changes, can be reduced through the continuous exchange of current design solutions between the involved development teams. The challenge is to find the optimum between the number of exchanges to reduce the negative effect of rework and the expense for the communication time. Only if concurrency and communication are simultaneously considered, the highest possible uncertainty reduction and hence, the optimal time-to-market is achieved.

The optimal method for generating such information during the NPD process depends on the development stage as well as on the type and degree of uncertainty. In the following, we will briefly survey the most important models and review the limited number of empirical studies that examine their respective effectiveness. As we will model the update of market requirement information in our decision framework, the focus will primarily be on methods reducing the market uncertainty. However, approaches to reduce other sources of uncertainty, in particular technical uncertainty, cannot independently be treated of each other and will therefore be sketched first.

In the early phase of the development process, uncertainty regarding the feasibility of certain technical solutions can be reduced by increasing the application of simulation and other virtual development techniques, e.g., digital mock up (DMU). Dahan and Srinivasan (2000) show that virtual prototypes are nearly as effective for concept selection and testing as physical ones. Where the latter are required, one can use rapid prototyping instead in order to obtain a physical prototype from computer models. This concept has proven to be an effective means to embody product concepts quickly and inexpensively, which enhances technical problem solving during the early development stages (Wall et al. 1992). The optimal prototyping and testing strategy as well as the optimal switch from virtual to physical modes depends on the inherent uncertainty and the cost of redesign (Thomke 1998; Thomke and Bell 2001).

Such prototypes are also very valuable for communicating product concepts to potential customers at an early stage. Especially for very innovative products, early feedback is crucial for product success. The integration of so-called lead users is therefore valuable (cf. e.g., von Hippel 1986; Albach 1993, p. 275 ff.). Since these customers expect great benefits from the developed product, they are willing to support the development process by bringing in their knowledge and experience or by testing early prototypes. The value of the obtained information depends on the degree and the point in time of their involvement as well as the respective incentives for them. With respect to time, lead user integration seems to be most valuable during the early stages for the evaluation of product concepts as well as during the testing phase (Gruner and Homburg 2000). During the design phase, the involvement often causes disturbances and bears the risk of addressing the specific lead user needs too much, thus drifting towards a niche market solution (Brockhoff 1997b).

The most common means of market information generation are, however, the traditional market research methods (Lynn et al. 1999; Zahay et al. 2004). Besides still prevailing simple customer surveys, multi-attribute models like conjoint analysis are the most popular methods for the identification of customer requirements prior to the start of the development project. The latter methods are used to determine the relative importance of certain product features by asking customers to evaluate alternative product concepts characterized by a set of attributes. The obtained data allows to estimate the potential market share of each concept in this set of alternatives and to derive an optimal product concept. In a similar way, quality function deployment (QFD) links customer needs to design attributes (cf. e.g., Griffin and Hauser 1993; Brockhoff 1999, p. 178 ff.). This method differs from the simple comparison of concept alternatives based on multi-attribute insights. QFD particularly fosters the interaction between marketing and engineering during the process of converting customer needs into engineering solutions. Thus, it also has an organizational impact.

The problem of either method is that the customer requirements are surveyed prior to the start of development activities and that they are generally not updated (von Hippel 1992). Thus, problems arise if the needs change. This is especially the case in highly dynamic environments (Bhattacharya et al. 1998). In addition, customers often do not know themselves which products they will need at the moment of the market launch. Although these are known problems, hardly any study exists measuring the accuracy of such market studies or the reliability of the applied methods. Among the few, Mahajan and Wind (1992) empirically study the usage and satisfaction of frequently applied market research methods in NPD projects. They report inaccuracy (e.g., of QFD, focus groups, and life cycle models) and inability to capture the market complexity (e.g., of conjoint analysis or focus groups) as the major shortcomings of these models. Based on a sample of 168 firms, Kahn (2002) reports an average accuracy for market and customer forecasts of about 58% over an average forecast time horizon of 26 month.[2] For products with a higher degree of innovation, the forecast accuracy is even lower, i.e., 47% for new-to-the-company and 40% for new-to-the-world products.

Other forecast accuracy studies with a particular focus on NPD projects seem only to exist for sales and profit forecasts. However, these studies are rare and partly outdated (Gartner and Thomas 1993). Tull (1967) as well as Tull and Rutemiller (1968), for example, compare the actual and predicted sales for new products based on a sample of 53 products from 16 firms. They report high inaccuracy for sales as well as profit forecasts with an average mean error of 65% and 128%, respectively. Beardsley and Mansfield (1978) also find a relatively low correlation of 0.37 between initially forecasted and actual profits of 57 product and process innovations of a single firm. Finally, Shelley and Wheeler (1991) report an average ratio between actual and forecasted sales of 79% in the first year which continuously decreases to 41% in the fifth year.

Although these studies measure only the accuracy of sales forecasts made at the moment of the product launch, the accuracy of forecasts made prior to the start of the project or during the development process might be even worse. Together with the findings reported above, these studies clearly indicate the need for regular updates in order to reduce market uncertainty for optimal decisions during the development process. A positive impact of information updates on NPD success has also been empirically validated (cf. e.g., Rothwell et al. 1974; Balbontin et al. 1999). Despite these insights, the number of corresponding models is limited and has not been of major interest in disciplines like management science or operations management (cf. e.g., Wind and Mahajan 1997; Krishnan and Loch 2005). The development of more formal and quantitative models in order to improve the resolution of uncertainty and to enhance decision making especially during the early stages is therefore a frequent request of researchers in this field (Mahajan and Wind 1992; Gerwin and Susman 1996). We will thus review the existing information updating and valuation approaches and evaluate their applicability to this problem in the next sections.

2.2 Decision Models with Information Updating

Researchers in various disciplines have developed decision models that deal with information updating. In the following, we will provide a brief overview of information updating within the decision theory before surveying the most interesting models related to our research objective. Since

---

[2]Unfortunately, the term "forecast accuracy" is not precisely defined in this article. It is only stated that the participating firms were asked to indicate the degree of forecast accuracy one year after the product launch.

information updating has hardly been addressed in the area of NPD, we will study similar decision frameworks from related disciplines, like supply chain management for example, and present the most relevant updating approaches. The focus of the latter aspect will primarily be on statistical decision theory, in particular Bayesian analysis, which we will apply to our valuation model. Finally, some decision models in the field of NPD that incorporate information updating will be analyzed in greater detail.

2.2.1 Foundations of Decision Theory and Information Updating

Decision theory is the study of decision making involving and being of interest to researches in many different disciplines like mathematics, statistics, economics, philosophy, psychology or behavioral science – to name but the most important ones. The vast body of knowledge goes back as far as to the eighteenth-century like to the theory on the measurement of risk by Daniel Bernoulli (1738) or to the notion of probability as a theory of rational degrees of belief developed by Thomas Bayes (1764).[3] The modern decision theory builds up on the work of von Neumann and Morgenstern (1944) and Savage (1954), who provide with the expected utility theory an important foundation of decision making under risk. One part of it – the so-called prescriptive decision theory – is concerned with the derivation of optimal strategies when a decision maker is faced with several decision alternatives and an uncertain or risk-filled pattern of future events (Laux 1991, p. 3 ff.).

An important characteristic of many decision problems of this type is that the decision maker's uncertainty is not constant over time. It rather depends on the point in time when the decision is made. As time progresses and the moment of the uncertainty resolution comes closer, the decision maker is generally less uncertain than he was at times farther away. The reason for the uncertainty reduction is that he can acquire additional information and thus, learn about future states of the world as time goes by (Marschak and Nelson 1962). This property is at the heart of information updating. The additional information is, however, only valuable for the decision maker if he has flexibility to respond to it (Merkhofer 1977). Numerous models for decision making in the presence of managerial flexibility to respond to new information in uncertain environments have been developed. NPD projects are characteristic examples where management has to refine its information over time and adjust its initial decisions.

The basic idea behind these decision models under preliminary information is the following (see also Fig. 2.2): In situations with uncertainty, the decision maker has often prior information about the unknown states of

---

[3]The latter seminal work builds a cornerstone in the statistical decision theory to which we will refer to in our model formulation. See Section 3.2 for details.

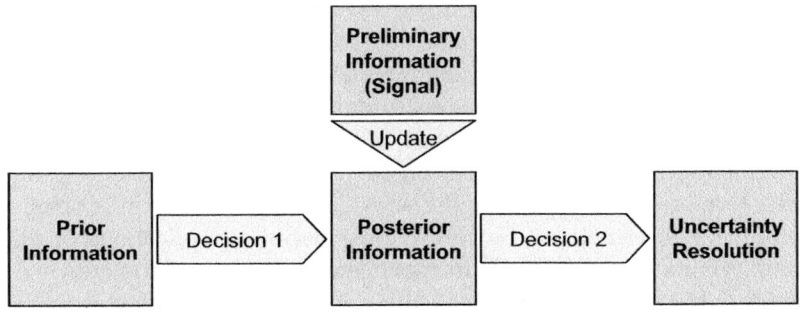

Fig. 2.2. Decision making with information updating

nature. In order to make the best possible decision, however, he may have the flexibility to postpone (part of) the decision until additional information becomes available. This new information can then be used to update the initial (prior) probability estimates about the state of nature so that the final decision is based upon more accurate data. The postponement of some decisions until a later point in time when more information is available generally comes at the expense of higher second stage costs. Late decisions that require fast production modes or design changes close to the market launch incur much higher costs compared to early actions. Thus, a frequent trade-off is to make either an early decision at low cost but with limited information or to postpone it until more information becomes available at the expense of higher cost (Loch and Terwiesch 2005).

Although many decisions in operations management involve this issue, corresponding models have been developed so far primarily in the area of supply chain management where information updating is a prevailing issue.[4] Key decision problems with uncertainty reductions through information or forecast updates in this fields are, for example, optimal inventory replenishing policies (e.g., Johnson and Thompson 1975; Lovejoy 1990), information-sharing mechanisms to reduce variability effects in supply chains like the bullwhip effect[5] (e.g., Lee et al. 2000; Gaur et al. 2005), or optimal ordering and production modes for items with uncertain demand patterns (e.g., Hausman and Peterson 1972; Fisher and Raman 1996; Eppen and Iyer 1997). The challenge underlying these decision problems is

---

[4]For a systematic overview of the development of dynamic inventory modeling under uncertainty and the corresponding mathematical and statistical methods of it, see, for example, Girlich and Chikan (2001).

[5]The bullwhip effect describes the phenomenon that the sequence of order quantities tends to have a higher variability and a larger order size as one moves upstream the supply chain, i.e., from the retailer to the manufacturer (Lee et al. 1997).

the optimal response to high demand variability. Depending on the studied issue, the demand uncertainty and its reduction is hereby modeled in different ways. Three major updating methods applied to decision models can be distinguished: time series analysis, Markov-modulated forecast updates, and Bayesian analysis (cf. Sethi et al. 2005, p. 8 ff.).[6]

In the following, we will focus on the latter method which is – compared to the former – best suited for the above described decision problem with preliminary information.[7] Since the most popular applications of this updating method are supply chain management decision problems, we have to refer to the models in this area. Although the context differs from the one of NPD projects, the underlying decision problem is identical: Instead of demand uncertainty, management of NPD projects has market uncertainty that can be reduced with new information acquired during the development process. The obtained insights from the supply chain management literature can therefore be applied to our decision problem. However, as the subsequent discussion will show, the applied methodology cannot be transferred in a straightforward manner, but has to be slightly adjusted.

2.2.2 Bayesian Analysis

2.2.2.1 Foundations and Basic Ideas

Bayesian analysis is a popular method in the field of statistical decision theory which is concerned with the problem of making decisions based on statistical knowledge about uncertain quantities. In order to obtain information about critical parameters, the decision maker faces the challenge of designing appropriate studies, analyzing data sets, and fitting probability models. While classical or frequentist statistics uses the sample data from experiments or studies directly in order to make inferences about the unknown parameters, Bayesian statistics[8] combines this data with other relevant information about the problem. This so-called prior information primarily arises from other sources than statistical investigation, like past project experience, for example. By combining the prior information about the states of nature of the decision problem with the additional information obtained

---

[6]These very general forecasting methods are also known in the literature as the state space forecasting approach (cf. e.g., Abraham and Ledolter 1983, p. 359 ff.).

[7]For the characteristics and shortcomings of time series analysis see e.g., Veinott (1965), Kahn (1987), or Lee et al. (2000) and of Markov-modulated forecast updates see e.g., Chen and Song (2001), Sethi et al. (2005), or Karr (1991).

[8]Named after Thomas Bayes, a minister and amateur mathematician, who set down his findings on probability in an "Essay Towards Solving a Problem in the Doctrine of Chances", published posthumously in the Philosophical Transactions of the Royal Society of London (Bayes 1764).

from (market) research or experimentation, the posterior distribution, i.e., the conditional distribution of the unknown quantities given the additional data, can be computed. All further inferences are then made from these updated beliefs (cf. e.g., Berger 1985).

In the past, there has been a long and intense controversy between frequentists and Bayesians statisticians regarding the appropriate approach to statistical data analysis and decision making (Berger 1985, p. 124 f.). The former group developed methods based on sampling from a large population which dominated the field for a long time. They criticized the Bayesian methods for their high dependency on correct and robust priors, the often subjective estimates, and their over-reliance on computationally convenient priors. The latter criticism refers to the fact that previously posterior distributions could only conveniently be determined for models where the prior belongs to the same distributional family as the sample data (so-called conjugate families).[9] Bayesians in turn complained that the frequentist approach ignores to incorporate relevant and insightful prior information and that this method requires large samples to derive significant and robust results which is often costly and inefficient.

Most of these controversial issues have been overcome. Computational advances from the mid-1980s on allowed to apply simulation methods, in particular Markov chain Monte Carlo simulations, to determine posterior distributions for a wide class of distributions and models (cf. Rossi et al. 2005, p. 1). Although frequentism remains to be the more robust approach due to its reliance on larger sample data, it is less suited for making decisions on the basis of limited information. More precisely, many classical large sample procedures simply fail if only a small data basis is available. Bayesian procedures with their ability to combine the sample data with prior estimates or beliefs generally provide better results in such situations and hence, would almost always be preferable (cf. Berger 1985, p. 125). The latter aspect is important for many real-life situations where the available data basis is often sparse due to budget or time constraints.

In addition, Bayesian analysis follows the natural way of making decisions in practical situations by starting with subjective estimates about uncertain outcomes of the project which are later revised and updated when new information becomes available. The derived insights and decisions from such an analysis are also more easily interpretable by non specialists.[10] The superiority in many situations explains the dramatic increase in the use of Bayesian methods in the different academic disciplines over the

---

[9]For details, see Section 3.2.1 or Carlin and Louis (2000, p. 25 ff.).

[10]Besides these benefits, there exist several other advantages Bayesian analysis offers compared to the classical statistical methods for which the interested reader is referred to the excellent textbook of Berger (1985, p. 124 f.).

last decade. In the following, we will review prominent applications of this methods in the field of supply chain management, real options analysis, and new product development.

2.2.2.2 Applications in Supply Chain Management Literature

Bayesian analysis has probably been the first updating method applied to the above discussed inventory and supply chain problems with uncertain demand patterns. Dvoretzky et al. (1952) were the first who studied Bayesian models to learn about future demand by combining prior distribution with additional information. Scarf (1959, 1960) derives an adaptive optimal order policy depending on the past history for the case of exponential demand distributions. Other contributions to this stream of research have been developed, for example, by Iglehart (1964) or Waldmann (1979) who are mainly concerned with deriving and characterizing optimal inventory replenishment policies. Noteworthy is the work of Azoury (1985) who models a periodic review inventory problem with several unknown parameters of the demand distribution as a Bayesian dynamic program with a multidimensional state variable.[11] The still prevailing problem by that time of computational intractability limited the attractiveness of Bayesian analysis. Non-Bayesian formulation of inventory problems as an approximation to the Bayesian dynamic programs turned out to be not equivalent as Azoury and Miller (1984) show. More recently, the interest in Bayesian analysis increased again. Lariviere and Porteus (1999) examine an empirical Bayesian inventory problem and derive optimal policies for both single and multiple market settings while Lovejoy (1990) studies exponentially smoothed forecast updates as well as Bayesian updates for myopic inventory policies.

Besides these Bayesian demand updates for the presented inventory problems, Bayesian analysis received special attention over the last decade in the Quick Response movement. Since the problem structure of these models is quite similar to ours of making preliminary decisions in NPD projects, we will have a closer look on these models. QR is an initiative of the apparel industry with the intention to reduce inventory costs in the supply chain by cutting manufacturing and distribution lead times through means such as better information exchanges between the participants, logistics improvements, and improved manufacturing methods.[12] The related decision problem corresponds to the above mentioned one of making decision under

---

[11] By considering distributions such as the gamma, uniform, Weibull, and normal, she extends the work of Scarf (1959) to other common classes of distributions where the known prior distribution is chosen from the corresponding natural conjugate family.

[12] For further details of this movement, the interested reader is referred to the excellent overviews of Hammond (1990) or Hunter (1990).

preliminary information: A buyer (e.g., a fashion retailer) places orders to a manufacturer over certain quantities before the actual demand is known. The retailer's trade-off of ordering too little, with the result of product stock-outs and low service levels (lost profit), or too much, with the result of increased holding costs and forced markdowns (additional costs), corresponds to the problem of the classical "newsboy" model[13] (cf. e.g., Nahmias 1997).

In the presence of long lead times and high demand uncertainties, as they are characteristic for the fashion industry for example, the (ex post) optimal order quantity is hardly ever met. To improve the decision making under such demand uncertainties, the classical newsboy model is extended for the QR environment by incorporating information updates of the initial demand estimates (Iyer and Bergen 1997). This allows the participants in the supply chain to delay the final quantity commitment until a later point in time when additional information becomes available. More precisely, the buyer can order some portion of the overall demand based on his initial estimates far ahead of the selling season. The costs of these items are relatively low as the production can be precisely planned and the long lead time allows to manufacture in low-wage countries. These prior estimates are then updated at a later point in time with data observed from related items, requests from trade fairs (e.g., fashion shows), or first orders received. Based on the adjusted demand estimates, a second order quantity can be placed for the remaining selling period if required. However, this quantity generally causes higher costs since the manufacturer has to resort to fast production modes, produce in plants close to the target market, or use a faster, but also more expensive conveyance (Kim 2003).

The developed models addressing the described decision problem provide insights into optimal order strategies (Fisher and Raman 1996; Eppen and Iyer 1997), e.g., the optimal order quantity at the different decision points, or show the benefits of QR for the different supply chain members (Iyer and Bergen 1997). Given our research objective, we are, however, solely interested in the applied updating mechanism. As indicated before, the update of the demand in these decision models is generally modeled in a Bayesian manner. Two major types of formulations can be found. The first group of authors assume that the initial and the total demand of a certain product is bivariate normal distributed (cf. e.g., Fisher and Raman 1996; Kim 2003; Gurnani and Tang 1999). This is one of the simplest distributional choices often made in Bayesian analysis since it allows to determine relative simple expressions for the moments (i.e., mean and variance) of the updated (posterior) demand distribution. The drawback is, however, that

---

[13]Also referred to as the newsvendor or newsperson model by overly politically correct persons. We take the latitude to ignore such nonsense and retain the original nomenclature.

such a distributional relationship has to be empirically validated. Without at least some indications of a bivariate normal relationship between the initial and total demand, the derived results have no practical relevance. Some authors like Kim (2003) or Gurnani and Tang (1999) ignore this prerequisite. They only take the benefits of the computationally convenient updating formulation without proving its applicability. Although the estimation and validation of demand densities and the corresponding parameters is a challenging task (Fisher and Raman 1996), it is the only way to justify the choice of this formulation.

For all other cases, where such a simplification is not applicable, one can still rely on commonly known distributions by applying the above mentioned concept of conjugate families where a prior distribution is chosen that is conjugate to the corresponding likelihood function. This updating formulation can be found, for example, in the QR models of Iyer and Bergen (1997) or Eppen and Iyer (1997). Although the determination of the posterior distribution is mathematically more complex, it is the only approach to develop an insightful model with practical relevance when the empirical validation of the chosen distribution is omitted or corresponding data is sparse. As the next section will show, the few R&D decision models of this type rely therefore on the concept of conjugate families.

2.2.3 Updating Mechanisms in R&D Models

The presented concepts of information updating have – albeit with significantly less effort – also been applied to decision problems in the field of R&D. In their influential work, McCardle (1985) and Lippman and McCardle (1987) develop a stopping model for management who faces the decision problem of adopting an innovative technology. Since the profitability of a new technology is quite uncertain at the moment of its announcement, management has to estimate its expected returns. However, prior to making the decision whether or not to adopt the technology, the firm has the possibility to sequentially gather additional information and to use it to update its initial profitability estimates. The update of the expected returns is modeled in a Bayesian manner assuming a conjugate relationship between the firm's prior distribution about the economic value of the technology and the distribution from which the information is generated.[14]

The model provides a clear policy for the information updating process. As soon as one of the two identified thresholds is crossed, the firm should stop the collection of additional information. In case that the upper limit is

---

[14]McCardle (1985) studies information structures which rely on the following conjugate families: Beta-Bernoulli, Gamma-Poisson, Gamma-exponential, and normal-normal, where the mean is unknown.

met, management should adopt the new technology, while the lower one indicates to reject it. Lippman and McCardle show that a higher precision of the prior distribution shortens the decision to adopt an innovation. In addition, better information (a sharper signal) only leads to the same effect if the precision of the signal is higher than the one of the prior distribution. Otherwise, it results in a delay of the decision.

These results are also studied in the presence of competition in a game-theoretic extension of this model provided by Mamer and McCardle (1987) who show that an increased expected level of substitutive competition reduces the probability that the firm will adopt the technology. Finally, Lippman and McCardle (1991) extend the sources of uncertainty in the model by adding to the uncertainty about the economic viability of a known technology the emergence of a new and unknown technology during the decision process. In a slightly different context, Krishnan and Bhattacharya (2002) focus on the role of design flexibility in the technology selection process and compare different design approaches (parallel path versus sufficient design).

Other models that address the management of preliminary information in NPD projects focus more on the exchange or the management of such information, e.g., Bhattacharya et al. (1998) who study the timing of product definition in highly dynamic environments where uncertainty is resolved through frequent, repeated interactions with customers. Since all these frameworks have in common that they do not focus on the update of information in a particular manner, but model preliminary information as a possible set of design parameters (Sobek et al. 1999; Krishnan et al. 1997) or a stream of engineering changes (Loch and Terwiesch 1998; Ha and Porteus 1995), they are not of interest to our research objective and hence, not further reviewed.

The reason for the few applications of information updating models to NPD problems lies – compared to the above discussed supply chain management models – in the complexity of the development projects and the underlying information structures. Loch and Terwiesch (2005) claim that the topology of the decision and outcome spaces underlying the currently existing decision models in the field operation management is not suited for NPD settings. Whereas in the supply chain management models, like the Quick Response models for example, the decision (how much to order) and the outcome (realized demand) correspond to a one-dimensional, ordered decision space, the information exchange in a NPD environment is more complex. In a product development project for example, where concurrent engineering is applied, multiple interdependent development activities are performed in parallel in order to shorten lead time (e.g. Krishnan et al. 1997; Loch and Terwiesch 1998; Terwiesch et al. 2002). Thus, information about several design specifications have to be exchanged simultaneously, each consisting of a set of multiple possible outcomes, i.e., the information structure and the decision space are multidimensional.

Loch and Terwiesch (2005) therefore use in their insightful decision model based on preliminary information the concept of information structures. In contrast to the one-dimensional structures of the currently existing Bayesian updating models, these information structures consist of set-based probabilities – formally represented by a sigma field – that are refined over time. In other words, the preliminary information is represented as a space of relevant outcomes where the aggregated events of incomplete information become known over time. The costs of the actions are assumed to increase over time. Modeling the decision problem as a two-stage stochastic dynamic program, the authors are able to derive optimal policies in dependence on the underlying cost structure as well as the available information gain. They show that waiting for more information and avoiding actions in an early period is the optimal policy if the cost increase is moderate. However, if building costs are so high that it is worth to delay actions while cancellation costs are not too high, the optimal managerial policy is iteration, i.e., take more targeted actions early and then adapt when new information becomes available. Finally hedging, i.e., starting with several actions simultaneously, is the optimal strategy if early actions are cheap compared to late ones.

With this decision model, Loch and Terwiesch provide one of the first quantitative frameworks for decision making based on preliminary information that applies to NPD projects. Although this model demonstrates the impact of cost and information characteristics on the optimal decision, it is less suited, as the authors claim themselves, to quantitatively evaluate managerial decisions. In addition, it does not allow to explicitly determine the value of additional information. But especially the latter aspect is of particular interest in NPD projects as it enables the assessment of the uncertainty reduction through an information update as well as its impact on the overall project value in financial terms. In the next section, we will therefore review financial methods that provide means for such a valuation.

2.3 Real Options Analysis

One key success factor of new development projects is, as the discussion in the previous section has shown, to understand the underlying uncertainty and respond to it accordingly. Although investments in such projects are generally irreversible, management often has the possibility to adjust its course of action during the development process as the uncertainty is gradually resolved with the arrival of new information. This flexibility to respond contingent upon additional information in such investment projects represents a value for the company that is neglected by the traditional investment valuation methods. Real options theory, by contrast, takes this operational flexibility in the valuation of such projects explicitly into account and is therefore well suited for our research objective.

In the following, we will briefly discuss the shortcomings of the traditional valuation approaches before giving an overview of the basic characteristics of real options and the corresponding valuation methods. Afterwards, we review key real options frameworks with applications to R&D projects and present first attempts of combining information updating with real options analysis. The focus will be on the concepts and contributions most relevant for our problem setting. For an in-depth survey of this growing area, see the comprehensive review of Lander and Pinches (1998) or the textbooks of Trigeorgis (1996) and Amram and Kulatilaka (1999).

2.3.1 Shortcomings of Traditional Valuation Methods

Most companies base their investment decisions on the results of traditional discounted cash flow analyses (Newton et al. 1996; Graham and Harvey 2001). These methods, like the net present value (NPV) or internal rate of return analysis for example, make implicit assumptions about the investment project under consideration. Since these measures require precise estimates about the generally uncertain future payoffs of the project in order to determine its value, they implicitly assume that the estimated revenues will actually occur. Moreover, they value the project solely on a go/no-go basis, i.e., the project can either be conducted now or never. Thus, they neglect the possibility to postpone the project for a certain period until additional information becomes available and (some) uncertainty is resolved. For these reasons, the discounted-cash-flow methods are not well suited for valuing NPD projects (cf. Kulatilaka and Marcus 1992; Haley and Goldberg 1995; Baecker et al. 2003).

Incorporating risk in the NPV analysis by adjusting the discount rate accordingly does not address the issue of imperfect cash-flow forecasts. Companies therefore frequently apply additional analysis methods like sensitivity analysis, traditional simulation, or scenario analysis. Sensitivity analysis allows to identify the key variables determining the cash flows and hence, their impact on the NPV. By analyzing the relative importance of a variable compared to the other ones, this analysis method indicates the riskiness of the different parameters and the potential impact of a misestimation on the project success. It studies, however, only the effect of one variable on the project value at a time while holding the other variables constant. Thus, it ignores interdependencies between the different variables (cf. e.g., Trigeorgis 1996, p. 52 f.).

This deficiency can be addressed by traditional simulation techniques, like Monte Carlo simulation. Instead of varying the primary variables determining the NPV once at a time, simulation methods specify the different variables with probability distributions obtained from empirical studies or subjective estimates and describe their interdependencies as well as their impact on the project value through a mathematical model. By using large-scale random samples from the probability distributions of all critical variables, a probability distribution of the NPV for the assumed investment strategy is obtained. The characteristics of the decision problem under uncertainty and the interdependencies of the different input variables can thus be captured and analyzed. One has to be aware, however, that it is generally very difficult to describe all project-inherent interdependencies of the different variables exhaustively and to precisely estimate their probabilities a priori. In addition, one has to be careful not to double-count for risk if the NPV is already determined based on a risk adjusted discount rate (Myers 1976). Moreover, as a forward-looking technique, simulation analyzes an a priori specified investment strategy, thus ignoring the managerial flexibility to adjust the preconceived decisions to upcoming contingencies.

Finally, scenario analysis is another frequently applied method to capture the underlying uncertainty of an investment. It studies possible future events by considering possible alternative outcomes. Most commonly, three different cases are studied, e.g., the most likely one as well as a worst and a best case scenario. The consideration of extreme cases, which is generally only insufficiently addressed in traditional simulations techniques due to the low probabilities of extreme values, improves the simple NPV analysis. By allowing for a more complete consideration of outcomes and their implications, the obtained insights from a scenario analysis improve the basis for investment decisions.

Although each of the just presented approaches accounts in its particular way for the project inherent uncertainty, as it is particularly characteristic for new product development projects, neither one incorporates the managerial operating flexibility to respond to contingencies or additional information. In other words, they all ignore the value to adapt the initial strategy contingent on the possible states of nature (outcomes) in order to capitalize favorable future opportunities, e.g., to defer, expand, contract, or abandon the project. Thus, they systematically underestimate the true value of the investment. However, for the correct valuation of a project, these existing real options have to be taken into account. Due to their close analogy to options on financial assets, the corresponding valuation methods developed in this area can be applied to determine the actual value of the project. Before describing these methods in greater detail, a brief characterization of the different real options will be given.

2.3.2 Real Options

An option is the right, but not the obligation, to take a certain action in the future contingent on the realized state of nature. It is thus valuable in the presence of uncertainty. In finance, a call option gives the holder the right, with no obligation, to acquire the underlying asset (with a current value V) for an a priori specified price (the strike or exercise price X) on or before a certain date (maturity date T). Similarly, a put option gives the right to sell the underlying asset for the exercise price. If the option can be exercised only at maturity by paying the option price C (for a call) or P (for a put), this type of option is referred to as a European option, while an American option also allows for an execution before maturity.

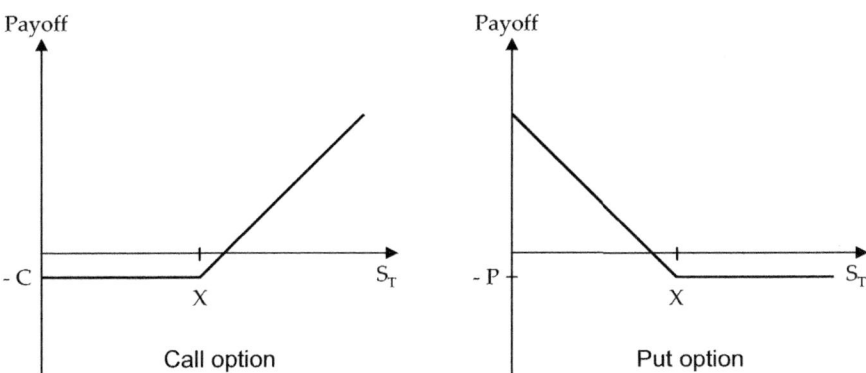

Source: Adapted from Hull (2003).

Fig. 2.3. Payoff pattern of a European call and put option at maturity

The benefit of an option results from its asymmetric payoff function (see Fig. 2.3). The reason of this one-sided payoff structure is that one will only exercise the option if it yields for the at maturity realized state of nature a positive payoff for its holder, i.e., a call option will be exercised only if the price of the underlying asset on that date exceeds the exercise price. Hence, an option allows to utilize the upside potential while limiting the downside risk.[15] If the underlying asset of the option is not a financial security, but the gross project value of discounted expected cash-inflows from investments in

---

[15]Contrary to options, financial contracts (e.g., forwards) have a symmetric payoff structure as they involve a commitment to fulfill an obligation undertaken to buy or sell an asset in the future at the terms previously agreed upon, regardless of the development of the underlying asset (McDonald 2003, p. 21 ff.).

Table 2.1. Comparison between financial call option and real option

Real option on NPD project	Variable	Financial call option
Present value of expected cash flows	S_0	Underlying stock price
Investment costs	X	Exercise price
Expiration date of opportunity	T	Maturity
Project value uncertainty	σ	Volatility of underlying stock
Risk-free interest rate	r	Risk-free interest rate

Source: Adapted from Trigeorgis (1996, p. 125).

real assets, like manufacturing plants, oil field exploration, or machinery for example, these options are called *real options* (cf. Myers 1977, 1984; Kester 1984). They are in contrast to financial options not specified by a contract, but embedded in capital-investment opportunities and have therefore to be explicitly identified. Table 2.1 depicts the close analogies between a real option and a financial call option.

Among the many real options identified to date, the literature distinguishes the following basic types, classified primarily by the source of (managerial) flexibility (cf. e.g., Trigeorgis 1996; Copeland and Antikarov 2001):

- Option to defer: This option allows management to delay its commitment to an investment project, e.g., new manufacturing plant, until additional information becomes available which justifies the expenditures. It is therefore particularly valuable in settings where management has the possibility to reduce the uncertainty (e.g., market uncertainty) over time through learning (cf. e.g., McDonald and Siegel 1986; Ingersoll and Ross 1992).

- Option to expand: This option can be viewed as a call option on an existing project to acquire an additional part of the base-scale project or to extend it by a certain percentage. It will therefore only be exercised if the future market development turns out to be favorable. Since it allows the company to capitalize future growth opportunities, the option to expand is of particular strategic importance. Birge (2000) shows how the value of additional capacity can be expressed in terms of option value, which leads to an application in capacity-planning.

- Option to contract: In analogy to the option to expand, this option type allows management to reduce the scale of its production if the market conditions, e.g., demand, price, volume, etc., turn out to be less favorable than initially expected (cf. e.g., Pindyck 1988).

- Option to abandon: This option provides management with the possibility to terminate a project permanently in exchange for its salvage value (cf. e.g., Myers and Majd 1990). The execution of this option may be beneficial if the current or future payoffs of a project do not even compensate for the fixed costs anymore. It thus provides a form of insurance against project failure. One should consider, however, not only all costs the termination may incur, but also take the loss of valuable expertise, organizational capabilities, or market access into account. The option to abandon is found, for example, in capital-intensive industries or new product development projects with high market uncertainty.

- Option to switch: The option to switch allows a company to change its mode of operation. This includes, for example, the option to temporarily shut down production and restart it later when the market conditions have improved (cf. e.g., McDonald and Siegel 1985). If the demand or the price of a certain product changes, the firm may also have the possibility to switch between alternative outputs (product flexibility). In the presence of such volatility, it may be valuable to invest in a more expensive, but also more flexible production mode that allows to adjust the output in accordance to the market development (Van Mieghem 1998). Similarly, management may be able to adjust the input to maintain the same output (process flexibility). This flexibility cannot only be achieved through an appropriate technology, as Kulatilaka (1993) shows considering the operation flexibility of a dual-fuel industrial steam boiler, but also through a flexible production or supply chain network.

- Option to improve: Introduced by Huchzermeier and Loch (2001), the option to improve represents the managerial flexibility in a new product development setting to take corrective actions during the development process in order to improve the performance of the product. The practical relevance of this option has been shown, for example, by Santiago and Bifano (2005) who value a high-technology development project in the presence of technical uncertainties.

In real-life investment projects, some of these just described real options may lead to interrelated opportunities which contain themselves other real options, i.e., compound options (cf. e.g., Geske 1979). Following Trigeorgis (1996, p. 132 f.), one can distinguish interactions between options on the same underlying project (interproject compoundness) and interactions embracing several underlying assets (intraproject compoundness). An example of the former compoundness is an investment in a R&D project which provides at its completion the opportunity for subsequent product generations or related applications. In this case, the subsequent projects are interdependent on the first one. In other words, the initial R&D project may lead to a whole chain of interrelated projects which in turn will offer future growth opportunities (Childs et al. 1998).

A multi-stage project, where an earlier investment is the prerequisite for the acquisition of the subsequent option to continue or extend the project, is an example for an intraproject interaction. Another example would be a project with several real options that are, if combined, simultaneously upward-enhancing and downward-protective. Brennan and Schwartz (1985) were the first who addressed this issue by studying the operation of a mine with the options to shut down and restart or to abandon it for its salvage value. A similar compound option in the context of a NPD project has been analyzed by Huchzermeier and Loch (2001) as well as Santiago and Vakili (2005). They analyze a project with an improvement option to increase the technical product performance by investing in additional resources – thus enhancing the upside potential of the project – and a coexisting abandonment option that simultaneously provides a protection against the downside risk.

Note that such interactions may affect the value of the real option. Trigeorgis (1993) has shown that the existence of subsequent options leading to further opportunities may increase the value of the effective underlying asset for earlier options. This leads in case of multiple interacting options on the same underlying to the fact that the separate option values do not necessarily add up, i.e., the combined value of the different options may differ from the sum of their separate values – they are so-called subadditive. Since this issue does not occur in our model, we will not further elaborate on this aspect, but refer the interested reader to the discussion in Trigeorgis (1993) or Kulatilaka (1995).

2.3.3 Valuation of Real Options

Due to their close analogy to financial ones, real options are valued by applying the concepts developed for pricing financial options. The seminal work of Black and Scholes (1973) and Merton (1973) marked the breakthrough in the valuation of financial options by providing a closed-form solution. They showed that – in an arbitrage-free world[16] – the price of an option equates the cost for setting up a continuously traded portfolio of the underlying security and a risk-free bond that exactly replicates the payoff of the option. The resulting partial differential equation can be solved in closed form, which is now known as the Black-Scholes option pricing model. The value of a European call option C can be determined as follows:

$$C = SN(d_1) - Xe^{-rT}N(d_2), \tag{2.1}$$

---

[16]The no arbitrage principle is also known as the "law of one price" which states that two investments with identical payoffs at all times and in all states must have the same value (cf. Brealey and Myers 1996, p. 961 f.).

where $d_1 = (\ln(S/X) + (r + \sigma^2/2)T)/\sigma\sqrt{T}$, $d_2 = d_1 - \sigma\sqrt{T}$, and $N(\cdot)$ is the cumulative standard normal distribution function.[17] The underlying assumption of the pricing model is that the stock price follows a stochastic Markov process that can be described by a geometric Brownian motion (or Wiener process).

A simplified valuation approach for options in discrete time was developed by Cox et al. (1979) where the stock-price movements follow a multiplicative binomial process. This model has proved to be very powerful in handling complex options and gained particular relevance for the valuation of real options in staged investments, like NPD projects that typically follow a stage-gate process. Especially if the resulting lattice tree is recombining, the underlying stochastic process can be well illustrated and the valuation of the real options in a backward recursive manner is intuitively traceable even for non-specialists.

The contingent claims analysis builds upon the basic risk neutral argument of Cox and Ross (1976). At the heart of this idea lies the above mentioned recognition that an option can be replicated from a continuously adjusted portfolio of traded securities. Since the value of the tracking portfolio and the option are independent of the investors' risk preferences, the valuation of the option can thus proceed in a risk-neutral manner, i.e., the expected future payoffs, weighted with risk-neutral probabilities, can be discounted at the risk-free interest rate. In other words, to value the managerial flexibility of an investment project, management neither has to estimate the probability of the future revenues nor the expected rate of return as long as a tracking portfolio can be created that incorporates the risk and return trade-off.[18] This makes the real options approach on the one hand a powerful tool, but limits on the other hand its practical applicability.

In many real-life settings, these underlying assumptions of real options valuation do often not apply (e.g., Lander and Pinches 1998; Brockhoff 2000; Witt 2003). Contrary to the valuation of financial options where the price of the underlying security is known from efficient financial markets, real options valuation lacks a corresponding market. The present value of an investment project (real asset) is generally not traded and thus, depends on subjective estimates. Although Trigeorgis (1993) as well as Mason and Merton (1985) argue that a dynamic portfolio of traded securities, which has the identical risk characteristics as the non-traded underlying real asset in complete markets, is sufficient for the valuation of real options (i.e., the existence of a continuously trading opportunity of the underlying asset itself is not required), it still requires a perfect correlation of the risk structure.

---

[17]For the definition of the variables, see also Table 2.1.

[18]The price of the underlying asset and its volatility has of course to be determined.

However, investment projects in real assets, and NPD projects in particular, have often idiosyncratic risks that are uncorrelated with financial markets (Huchzermeier and Loch 2001). Thus, the by options theorists favored contingent claims analysis has limited applicability.

Decision scientists, on the other hand, rely on the alternative approaches of decision trees and stochastic dynamic programming models to determine the value of real options (cf. Bonini 1977; Dixit and Pindyck 1994, p. 93 ff.). These models capture the decision maker's beliefs about the outcome of the project by assessing subjective probabilities for the uncertainties while representing the preferences for project cash flows by a risk-adjusted discount rate. The latter aspect, however, raised major criticism by options theorists, e.g., Trigeorgis and Mason (1987) or Mason and Merton (1985), who claim that the problem of finding the correct discount rate remains unsolved in these models. Smith and Nau (1995) therefore propose a combined approach that results – when including market opportunities in the decision tree analysis and capturing time and risk preferences through a utility function – in the same project value and optimal strategy as contingent claims analysis. Unlike in the latter model, subjective beliefs and preferences play a critical role in this integrated valuation approach. For an application of this model to oil explorations projects see Smith and McCardle (1998).

Regardless of these two valuation philosophies with their particular advantages in specific settings on the one hand, but known shortcomings on the other hand, real options analysis in the broader sense is the appropriate method to determine the value of an investment project under uncertainty when management has the flexibility to respond to new information. It has therefore received increasing attention from academics as well as practitioners in the last decades (cf. e.g., Graham and Harvey 2001). Traditionally, many real options valuation models and practical applications can be found in the area of natural resource exploration projects, like oil drilling or mining projects, where the above mentioned difficulty regarding the replicating portfolio does not occur due to the existence of similar traded assets (e.g., Brennan and Schwartz 1985; Cortazar and Schwartz 1998; Cortazar et al. 2001; Kamrad and Ernst 2001).

Real options frameworks more directed towards operations management issues are less abundant. Kulatilaka (1988), for example, develops a stochastic dynamic programming model to value the flexibility in a manufacturing system stemming from the possibility to operate in different alternative modes. Kogut and Kulatilaka (1994) analyze the operating flexibility to shift production between two manufacturing plants located in different countries while Huchzermeier and Cohen (1996) focus on different manufacturing strategies exercised contingent upon exchange rate realizations in a global production network. Investments in flexible production capacity are valued by He and Pindyck (2002) who model this capacity choice problem as a singular stochastic control problem. For further references on real options frameworks focusing on product and operations management, see the comprehensive reviews of Lander and Pinches (1998) and Miller and Park (2002).

2.3.4 Models and Applications for R&D Projects

Besides the above mentioned application areas, real options valuation obtained special attention in the field of R&D since it is an ideal approach to value the managerial flexibility of responding to the high level of technology and market uncertainty inherent in these projects. In addition, the sequential development process of the projects fosters the application of the real options framework since it can be regarded as a series of investments, each containing one or more real options, like extending a project or switching to a different technology, for example. In the following, we will present some insightful frameworks that are of interest to our research objective.

Baldwin and Clark (1998, 2002) analyze the option value of modularity in product design. As such a design increases the flexibility to respond to (late) market requirement changes and makes the complexity in development projects manageable, it has substantial value. They show that the option value of a modular design approximately corresponds to the net option value inherent in each module less the cost of creating the modular architecture. Childs et al. (1998) develop a real options framework to study the optimal investment policy for product development projects that can either be developed in sequence or in parallel. Their analysis reveals that the optimal development strategy depends on the volatility, the correlation between the project's present values, as well as the development cost and time. Loch and Bode-Greuel (2001) focus on the complex sequential decisions in drug development projects and examine the inherent compound real options with a decision tree valuation framework.

Lint and Pennings (2001) use a real options approach to develop a framework that addresses market and technology uncertainty in a development project. They treat NPD as a sequential process in which management has the following options at the different stages: Extend the efforts to large scale R&D after the evaluation of the initial product idea screening, conduct R&D without the obligation of launching the project, invest in further validation of the product's market potential, and finally, launch the product into the market. With this framework, they compare R&D project portfolios containing these options that are exposed to different degrees of uncertainty. The already mentioned framework of Huchzermeier and Loch (2001) as well as the extension of Santiago and Vakili (2005) allow to evaluate flexibility in a single R&D project in the presence of compound real options. In contrast to

Lint and Pennings, they examine multiple sources of uncertainty (e.g., budget, schedule, technical performance, market requirements) and extensively address the distinction between these sources of operational variability and financial uncertainty.[19] As it allows to value not only the common option of abandoning the project, but also explicitly considers the possibility of taking corrective action in order to improve the technical performance and hence, the project payoff, this model is very realistic and insightful for valuing NPD projects. This makes it well suited for our decision problem. We will therefore provide a more thorough description of the model in Section 3.1.

Although these frameworks explicitly address the managerial flexibility to respond accordingly when uncertainty is resolved, they do not incorporate learning as an explicit element. The acquisition of new information is rather viewed as a passive consequence of the project progression. However, some approaches exist that address this issue. Childs and Triantis (1999), for example, study dynamic investment policies for a R&D program where management has the possibility to reduce uncertainty through investments in the development process. In particular, they model two R&D projects with a three-dimensional lattice tree and demonstrate how collateral learning between the projects as well as a dynamically altering funding policy (e.g., accelerating, shelving, or abandoning projects) affect the optimal investment policy.

Bellalah (2001) analyzes investment decisions under uncertainty and incomplete information with a continuous-time model. By explicitly accounting for information costs regarding the project cash flows, he is able to discuss the impact of these costs on the project value. Martzoukos and Trigeorgis (2001) explore the reduction of uncertainty in investment opportunities. By using a real options framework with incomplete information and costly learning actions that induce path-dependency, they show that optimal timing of information acquisition is essential as it reduces the cost of potential mistakes and hence, increases the value of investment opportunities. As additional information is generally costly, management has hereby to trade-off between the quality and the cost of learning.

The explicit incorporation of learning in these frameworks is predominantly modeled by considering corresponding costs for information. Thus, information acquisition is rather studied from a cost than from a decision making perspective. However, some attempts exist that combine statistical decision theory with real options analysis in order to incorporate information updates in a Bayesian manner. These models will be presented next.

---

[19]A practical application of this model to a high-tech NPD project is presented by Santiago and Bifano (2005).

2.3.5 Attempts of Combining Bayesian Analysis and Real Options

Herath and Park (2001) are one of the first who explore the idea of combining Bayesian analysis with real options. They develop a simple valuation framework based on the concept of the expected value of perfect information. With this approach, they study investment decisions where management has the option to defer a project until more information becomes available. The model allows for sequential revaluations of the project using sampling information in a Bayesian manner to reduce future uncertainty at each decision point.

Miller and Park (2005) build on this decision theory framework and develop a real options model that incorporates information acquisition through a Bayesian update. They model a contingent multi-stage investment scenario where management uses the information obtained after the first investment phase of the project to update the initial estimates of the expected future cash flows. Contingent on the adjusted estimates, the updated option value for the remaining project is determined and the decision about the next investment phase is made, i.e., whether to continue the project or not (other managerial options are not considered). Depending on the number of considered stages, this updating and decision procedure is repeated. Although the updating method applies the before mentioned normal conjugate relationship, it is one-dimensional and only allows for updates of the mean. The derived insights are therefore limited. The only key finding is a threshold which defines when the firm's prior decision is reversed based on the observed sample result.

A primarily practical application of a framework that combines Bayesian analysis with real options is presented by Armstrong et al. (2005). They study the option value of acquiring additional information for an oilfield production enhancement project. In this case, management has to value the investment in a workover of an oil well in order to maintain hydrocarbon production at a satisfactory economic level. Besides the choice between simply continuing the production and conducting a workover based on the currently existing information, management has also the possibility to increase the efficiency of the workover by conducting a study about the reservoir. The results obtained from this costly study can be used to update the initial information about the benefit of a workover. To determine the value of this additional information, the authors incorporate Bayesian analysis into a real options framework. More precisely, they assume that the two sources of uncertainty, the underlying oil prices and the characteristics of the reservoir, are bivariate normal distributed. Using Monte Carlo simulation to compute the option prices based on the assumed distributions, they are able to identify a threshold value for the oil price when the acquisition of the additional information is valuable.

These two recently published models are, to the best of our knowledge, the first attempts which combine Bayesian analysis with real options frameworks. Although both models are quite simple in their structure and provide only limited managerial implications, they clearly show that the combination of these two methods is a well-suited approach to value information updating.

2.4 Summary

The review of the relevant strands of literature in this chapter clearly indicates the need for more quantitative models to improve decision making in NPD projects that are inherently exposed to a high degree of uncertainty (see Fig. 2.4 for a summary). Although numerous empirical studies identify the timely generation of information (particularly in the early phases of the development process) as a key success factor and stress the need for regular updates, hardly any formal models exist. Most of the derived managerial insights are obtained either from normative or empirical studies. However, only quantitative models allow to determine the project specific optimal response to additional information by explicitly incorporating the underlying parameters, like development costs, time, expected payoffs, etc. In addition, they allow to study the impact of these parameters on the optimal solution and to determine the value of an information update during the development process.

The problem of making decisions under preliminary information is not limited to NPD. We have seen that other areas of operations management, like supply chain management for example, face similar challenges. The developed decision models in these fields apply various information updating methods to optimize operations by reducing the inherent uncertainty as time progresses. While time series methods or Markov models require sufficient historical data to extrapolate the future development or to model the underlying process of the unknown parameter(s), Bayesian updating allows to combine (subjective) prior estimates with additional information in a straightforward manner and hence, is best suited to model the uncertainty reduction in NPD projects.

Some first attempts exist which explicitly model information acquisition via Bayesian updating in a real options framework. By combining these two methods, one is able to determine the value of information acquisition. However, these rather simple models have neither a primary focus on NPD projects nor do they address the NPD specific characteristics like multidimensional information structures. Thus, an information updating model for NPD projects must address these characteristics as well as the typical sources of uncertainty. An insightful real options framework that allows to

Area	Insights for Thesis	Differences to Thesis	Key References
New Product Development	• Information generation and updating is a key success factor • Lack of decision/valuation models for early stages of NPD projects • Design flexibility in highly uncertain environments has high impact on project success	• Research studies are largely descriptive • Primarily qualitative managerial insights provided • Dependency of optimal decision on underlying parameters not addressed	• Mahajan (1992) • Thomke (1997) • Krishnan and Loch (2005)
Real Options Theory	• Well suited framework to value managerial flexibility in NPD • Integration of improvement option as a managerial action to increase expected performance • Only few attempts for integration of information updates exist	• Only one source of uncertainty explicitly modeled • No explicit determination of information updating value • Uncertainty reduction regarded as a passive process	• Huchzermeier and Loch (2001) • Santiago and Vakili (2005) • Miller and Park (2005)
Decision models with information updates	• General set-up to model decision making under uncertainty • Topology of outcome space has to address NPD characteristics • Integration of information updates with methods of Bayesian analysis	• Applied Bayesian updating models have only one-dimensional state space • Only mean updates considered • Insufficient focus on NPD problem structure	• Berger (1985) • Iyer and Bergen (1997) • Loch and Terwiesch (2005)

(Thesis)

Fig. 2.4. Summary of literature

value managerial flexibility in NPD project is the model of Huchzermeier and Loch (2001) and the extension by Santiago and Vakili (2005). It considers not only the in NPD widely applied option to abandon a project in case of an unfavorable technical or financial development, but also takes the possibility of corrective actions to improve the technical performance into account. This makes the framework well suited for NPD applications as well as for a base case of an information updating decision model. It will therefore be presented in greater detail in the next chapter.

3

Model Description

This chapter defines the information updating valuation model. Since our model builds upon an extended version of the before mentioned model for valuing managerial flexibility in R&D projects developed by Huchzermeier and Loch (2001), we will start with a brief description of it. In Section 3.2, we gradually derive the Bayesian updating formulation for an update of the mean and the variance of the market requirement distribution. This information updating framework is then integrated into a valuation model that allows to determine the value of an information update given managerial flexibility (Section 3.3).

3.1 Basic Model

Huchzermeier and Loch (2001) have developed an insightful decision model that allows to determine the value of managerial flexibility inherent in R&D projects. Santiago and Vakili (2005) have extended this model and derived some additional results. We will build upon this work by using a modified version of the former framework as the basis for our valuation model.

Our model differs in two aspects: Firstly, we will limit the product performance variability, which represents the technical uncertainty inherent in the development project. This allows us to reduce the complexity of the basic model without confining the practical applicability too much. In addition, we can derive key properties of the model in closed form which would be otherwise not feasible.[20] Secondly, while the authors of the former model basically assume that the future market success is determined by a single product performance parameter, we will – similar to Santiago and Bifano (2005) – also provide a model formulation for multiple performance parameters. This ensures the practical applicability of the framework

[20]See Appendix B for details of this assumption.

C. Artmann, *The Value of Information Updating in New Product*,
DOI: 10.1007/978-3-540-93833-0_3, © Springer-Verlag Berlin Heidelberg 2009

and simultaneously addresses criticism raised by other authors. Loch and Terwiesch (2005), for example, note that the major limitation of applying the existing Bayesian updating models of the supply chain management literature (cf. e.g., Fisher and Raman 1996; Iyer and Bergen 1997) to product development lies in the one-dimensional performance or outcome space.[21] Other environments, like NPD, are more complex since the performance and thus, the project success depends on multiple parameters. By also providing the multidimensional formulation of the model, we explicitly address this issue. In every other aspect, the structure of our basic model is identical with the one of Huchzermeier and Loch.

In the following, we provide a brief overview of our modified version of the Huchzermeier-Loch model, starting with the one-dimensional product performance case. We will hereby adopt the notation introduced by Santiago and Vakili (2005), which has some advantages for our later model formulation. For further details, the reader is referred to the respective papers.

3.1.1 General Structure and Development Uncertainty

The decision model of Huchzermeier and Loch assumes that the R&D project follows a classical stage-gate process where the project – initiated in $t = 0$ – is developed over a period of T discrete stages towards market introduction in $t = T$. The market success of the project depends on its realized performance which is measured by a one-dimensional parameter x.

During the development process, however, the project and hence, the product performance underlies uncertainty resulting from market and technical development risk which prevents to determine the project success with complete certainty upfront. Performance variability causes the product performance to move either up or down between the different stages of the project. This drift of the state variable x is modeled by a binomial distribution assuming that the performance state of the product depends on a random variable ω_t. The performance may either increase from period t to the next period $t + 1$ by $\omega_t = 0.5$ with probability p or decrease by $\omega_t = -0.5$ with probability $(1 - p)$ due to unexpected adverse events.[22] The performance change between the periods is hereby assumed to be independent

---

[21]See Section 2.2.3 for details.

[22]While Huchzermeier and Loch (2001) as well as Santiago and Vakili (2005) generalize the performance variability to be spread over N states (i.e., $\omega_t = \frac{i}{2}$ with probability $\frac{p}{N}$ and $\omega_t = -\frac{i}{2}$ with probability $\frac{1-p}{N}$ for $i = 1, \ldots, N$), we will limit it to one performance state. This constraint allows us to derive some general properties of our information updating valuation model in closed form. See Section 4.2.1.2 and Appendix B for details.

of the previous process history. Thus, the performance state of the development project at stage $t+1$, denoted by X_{t+1}, depends on the performance state at stage t plus the described development uncertainty ω_t, i.e.:

$$X_{t+1} = X_t + \omega_t. \tag{3.1}$$

We further denote a particular realization of the performance state X_t by x_t. The left hand side of Fig. 3.1 shows how the performance states of the project evolve over the different stages including the corresponding transition probabilities.

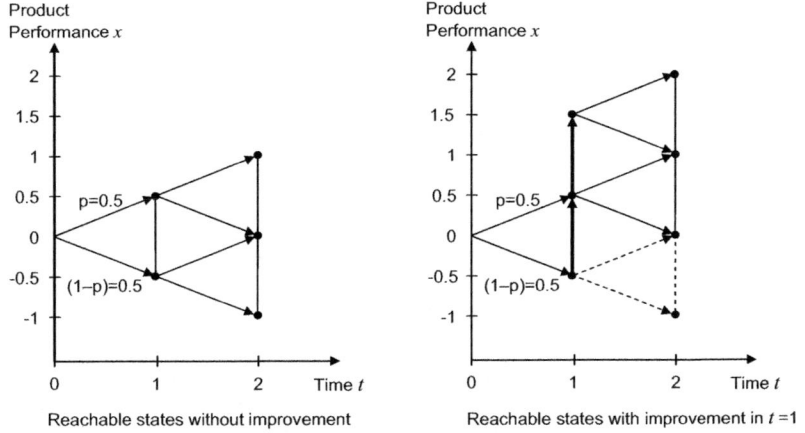

Fig. 3.1. One-dimensional product performance drift

3.1.2 Managerial Options and Development Costs

At each stage, management has the possibility to respond to the random performance change by choosing one of the following actions a_t: *continue*, *abandon*, or *improve*. The first action simply continues the project at a continuation cost of c_t; its performance in the next period depends on the random performance move. The option to abandon terminates the project at the current state, cutting all further costs as well as all future revenues.[23] Besides this standard real option, management has also the possibility to take corrective action. By investing in additional resources (e.g., additional engineers, outsourcing of design activities, etc.), the company can improve

---

[23]It is assumed that once a project is terminated, it will not be reactivated afterwards, i.e., it will remain in that abandonment state in all following stages.

the expected product performance, for example – and without loss of generality, – by one level (see right hand side of Fig. 3.1). This action incurs an improvement cost of α_t in addition to the continuation cost of the particular stage. Both managerial action costs, c_t as well as α_t, have to be paid at the beginning of each period.

Under consideration of this managerial flexibility, the performance state of the project at stage $t+1$ depends therefore on 1) the performance state at stage t, 2) the development uncertainty ω_t, and 3) on the management decision made at the beginning of stage t, a_t:

$$X_{t+1} = \begin{cases} X_t + k(a_t) + \omega_t, & \text{if } a_t = \text{continue or improve} \\ Stopped, & \text{if } a_t = \text{abandon} \end{cases}$$

where $k(a_t)$ represents the impact on the performance state X if action a_t is chosen at the beginning of stage t. It can take the following values: $k(\text{continue}) = 0$ and $k(\text{improve}) = 1$. The corresponding development costs can be summarized as follows:[24]

$$c(a_t) = \begin{cases} 0 & \text{if } a_t = \text{abandon} \\ c_t & \text{if } a_t = \text{continue} \\ c_t + \alpha_t & \text{if } a_t = \text{improve} \end{cases}$$

These costs have to be paid at the beginning of each period and are – like the revenues – discounted at the risk-free interest rate r. The rationale behind this assumption is that the risk of a NPD project is unsystematic which a rational investor can diversify away by holding a portfolio of securities without requiring a risk premium (Huchzermeier and Loch 2001). We follow their argumentation. As Smith and Nau (1995) have shown, the backward determination of the project value in a decision tree can therefore be done in a risk-neutral manner, i.e., using the risk-free interest rate.[25] In addition, an initial investment of I is required in $t = 0$ to the start the project (e.g., project infrastructure).

3.1.3 Market Uncertainty and Market Payoff

When the final product is launched to the market at time T with a product performance level of x, it will generate an expected market payoff, denoted

---

[24]In order to reduce complexity, we assume that the costs are only stage dependent. Performance state dependency could, however, be easily integrated, i.e., $c_t(x_t, a_t)$.

[25]See also Section 2.3.3 for a summary of the discussion regarding the appropriate discount rate in models using decision analysis versus contingent claims methods.

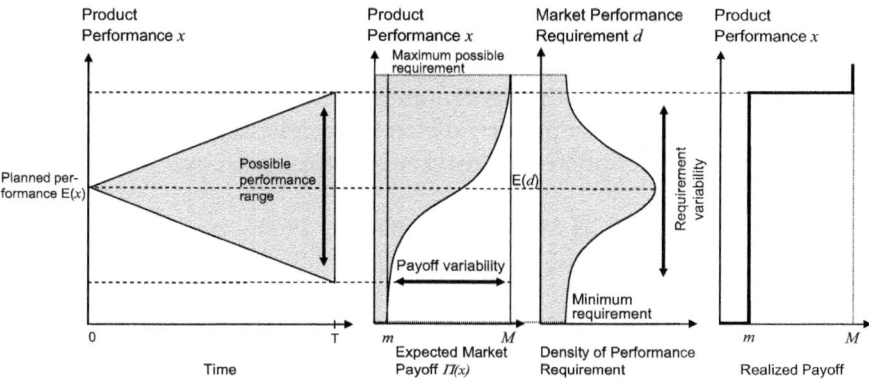

Source: Adapted from Huchzermeier and Loch (2001).

Fig. 3.2. Market requirement uncertainty and project payoffs

by $\Pi(x)$. In analogy to the general market payoff model, which has the form of an S-shaped curve, the expected payoff is assumed to be convex-concave increasing in x (Fig. 3.2). It is intuitive to assume that a performance improvement matters most for intermediate performance levels, while it has less impact when the performance is either very low or already very high (cf. e.g., Bhattacharya et al. 1998; Mahajan and Muller 1990).

It is further assumed that the payoff is the result of a competitive performance threshold which the firm does not know in advance. That is, the market requires a certain level of performance, denoted by the market performance requirement d. If the final product meets or exceeds the requirements of the market, the market will yield a high payoff M. On the other hand, if the final product performance misses the requirement target, it will only earn a much smaller margin m. The intuition behind this assumption is as follows: In the first case, the product has a competitive advantage in the market which ensures the company a price premium, while in the second case, the product is only one of many similar products and thus, has to compete on cost.

At the start of the project, the company does not know the required market performance level d. The management has only a forecast of d in form of a probability distribution which is obtained from market research and past project experience. We hereby assume that the market performance requirement level d is normally distributed. This seems to be reasonable since for many products the majority of the target customers have similar performance requirements, while a decreasing number of customers need either higher or lower performance requirements.[26] Thus, the density of the market performance requirement distribution has a single maximum at the mean $E(d)$ and the corresponding cumulative probability distribution is monotone increasing. The market requirement variability is represented by the variance of the market performance requirement distribution.

Under these assumptions, the market payoff is given by

$$P(x) = \begin{cases} M, & \text{with probability } F(x) \\ m, & \text{with probability } 1 - F(x) \end{cases}$$

where $F(x)$ represents the probability that the product performance x exceeds the market performance requirement d, i.e., $F(x) = P(d \leq x)$. The expected payoff is thus

$$\Pi(x) = E[P(x)] = m + F(x)(M - m). \tag{3.2}$$

It is further assumed that the targeted product specification (or expected product performance) at the start of the project is set in such a way that it meets the expected market performance requirement, i.e., $E(x) = E(d)$, which simplifies the exposition. In practice, management will probably set the product specification higher than the expected market performance, i.e., $E(x) > E(d)$, in order to have some buffer to meet or exceed the customer requirements. But note, the derived results do not depend on this assumption (cf. Huchzermeier and Loch 2001). Fig. 3.2 illustrates the market performance requirement uncertainty and the expected project payoffs of the model.

3.1.4 Value Function and Dynamic Program

The decision problem of the company at each stage is to select the managerial action that will reward the highest final market payoff. Since the variability of product performance, the cost of the different managerial options, as well as the expected market payoff are known, the firm can determine the value of the project with a stochastic dynamic program[27], which has the following value function:

$$V_t(x) = \max_{a_t} \left\{ -c_t(a_t) + \frac{1}{1+r} E_\omega \left[V_{t+1} \left(X_{t+1}(x, a_t, \omega_t) \right) \right] \right\}. \tag{3.3}$$

---

[26]This simplification, which is frequently assumed (cf. e.g., Kalyanaram and Krishnan 1997; Miller and Park 2005), provides some computational advantages in the subsequent information updating formulation. However, also any other reasonable distribution could have been chosen.

[27]See, for example, Bertsekas (2000) or Birge and Louveaux (1997) for details.

In other words, at the beginning of each stage t, the company compares the cost of the managerial action with its discounted expected payoff in $t+1$ and then selects the option with the highest expected return. $V_t(x)$ represents the real option value of the project, or short, project value in state x at stage t.

This stochastic dynamic program can be solved by a standard backward recursion which starts with the following terminal value at stage $t = T$:

$$V_T(x) = \Pi(x). \tag{3.4}$$

If $V_0(0)$, the optimal project value at stage $t = 0$, exceeds the initial investment cost I, the company would start the development project. Otherwise the project would be rejected. In the special case when "continue" is chosen in every decision point, the project value corresponds to the classical (or static) net present value. Note that the project value includes a compound real option, namely the value of the option to choose between improvement or abandonment in any period. Thus, the project value at time $t = 0$ consists of the NPV plus the option value (OV): $V_0(0) = NPV + OV$.[28]

3.1.5 Multidimensional State Space

The so far presented model considers only a one-dimensional performance state space which is sufficient for deriving properties for the value of managerial flexibility given the different sources of uncertainty in the development process. But it will fail in any realistic NPD setting where a project is defined by several technical parameters. Hence, in order to ensure the practical applicability of the framework, we will provide – similar to Santiago and Bifano (2005) – the model formulation for k performance parameters in this section. All previous assumptions will remain valid for the multidimensional case.

The performance state at the beginning of stage t is modeled by the vector $\boldsymbol{X_t}$ whose elements $X_1, \ldots, X_k$ represent the technical specifications or performance parameters of the project.[29] The performance parameters X_i are assumed to be independent. Without counter measures, the project may drift from period to period with the modeled randomness between the performance states in the particular dimensions. Thus, the performance state of the development project at stage $t+1$ depends on the performance state at stage t plus the described development uncertainty $\boldsymbol{w_t}$:

---

[28]The real options literature also refers to the option value sometimes as the option premium and to the real option value as the expanded NPV (cf. Trigeorgis 1996, p. 149). We will, however, use the expressions introduced above.

[29]From now on, vectors will appear in boldface type.

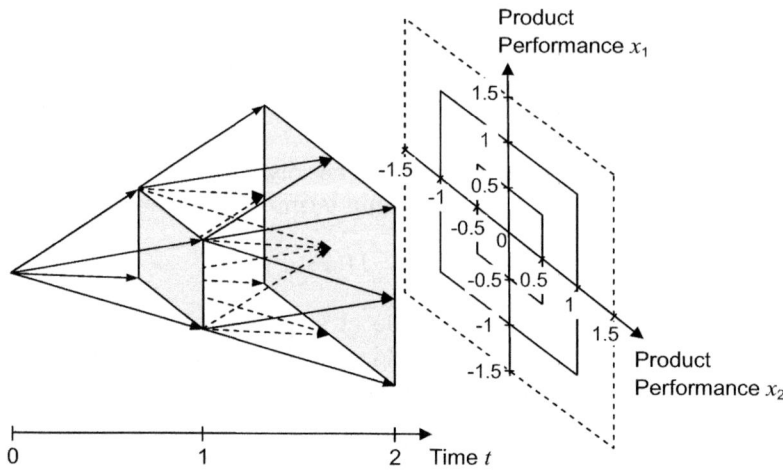

Fig. 3.3. Two-dimensional product performance drift

$$X_{t+1} = X_t + \omega_t. \tag{3.5}$$

Fig. 3.3 shows for a two-dimensional product performance vector how the performance state evolves over the different stages.

Management has the possibility to respond to this performance drift by either choosing continuation, abandonment, or improvement. The difference to the one-dimensional case is that improvement may include multiple options. For example, the firm could either improve the technical performance solely in dimension x_i or in any combination of the n dimensions (see Fig. 3.3). The improvement costs typically increase with the number of improved performance parameters. While it is conceivable that the expected improvement in dimension i could have an impact on the other dimensions j $(j \neq i)$[30], we will assume that any corrective action in dimension i is independent from any action in another dimension j.

If we denote all managerial decisions on dimension i of performance state X at the beginning of stage t, i.e., a_{it}, by vector a_t, we obtain the following relation for the performance state at development stage $t + 1$:

$$X_{t+1} = \begin{cases} X_t + k(a_t) + \omega_t, & \text{if } a_{it} = \text{continue or improve} \\ Stopped, & \text{if } a_{it} = \text{abandon} \end{cases}$$

---

[30]The following cases might be possible: 1) it might deteriorate the technical performance of the project in one or more dimensions; 2) it might have no impact; or 3) it might improve the performance. See Santiago and Bifano (2005) for details.

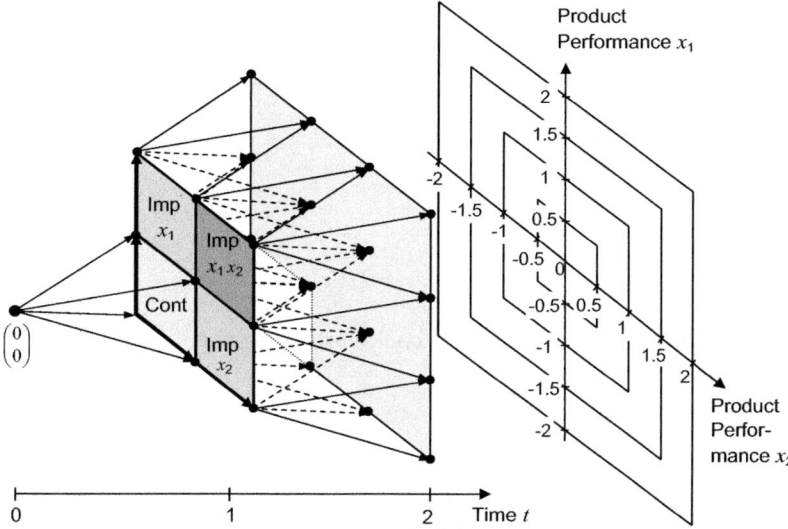

Fig. 3.4. Two-dimensional product performance drift under consideration of managerial options

where $k(a_{it} = \text{continue}) = 0$ and $k(a_{it} = \text{improve}) = 1$ corresponds the impact on dimension i of performance state X. Fig. 3.4 illustrates the possible performance states under consideration of the managerial options for a two-dimensional product performance vector.

The corresponding development costs can be summarized as follows:[31]

$$c(a_{it}) = \begin{cases} 0 & \text{if } a_{it} = \text{abandon} \\ c_{it} & \text{if } a_{it} = \text{continue} \\ c_{it} + \alpha_{it} & \text{if } a_{it} = \text{improve}. \end{cases}$$

The success of the project depends on whether the realized performance level $x = x_1, \ldots, x_k$ in the different dimensions meets the requirements of the market, denoted by d. With an efficient forecast of d in form of a multivariate probability distribution, the firm can determine the expected payoff as follows:

$$\Pi(x) = E[P(x)] = m + F(x)(M - m). \tag{3.6}$$

---

[31] As in the one-dimensional case, we also assume here that the costs are only stage dependent. Performance state dependency could, however, be easily integrated, i.e., $c_t(x_t, a_t)$.

Based on these assumptions, we can determine the project value in the multidimensional case by applying the known stochastic dynamic program with the following value function:

$$V_t(x) = \max_{a_t} \left\{ -c_t(a_t) + \frac{1}{1+r} E_\omega \left[V_{t+1} \left(X_{t+1}(x, a_t, \omega_t) \right) \right] \right\}, \qquad (3.7)$$

with terminal value at stage $t = T$:

$$V_T(x) = \Pi(x). \qquad (3.8)$$

It can be solved with the standard backward recursion.

3.1.6 Summary

The presented decision model of Huchzermeier and Loch (2001) values the managerial flexibility given an exogenous stochastic process of the technical uncertainty. At each of the T development stages, management has, besides simply continuing the project, the possibility of improving or abandoning the project when new information becomes available. The project value considers these managerial options which are neglected by the traditional NPV method.

In absence of perfect market information at the start of the project in $t = 0$, management has to make assumptions about the customers' performance requirements at the launch of the product in $t = T$. Although the resolution of external uncertainty in the chosen notation of Santiago and Vakili (2005) could be considered in ω (e.g., by letting the aggregate state of the project reflect the performance relative to the market requirement), an explicit update of the market performance requirement distribution as well as the explicit value of such an information update has not been modeled so far. In order to quantify the value of such (additional) information acquisition during the development process, we need to model this updating process with means of statistical decision theory.

Bayesian statistics combines initial beliefs with additional information acquired during the development process into posterior information. It is therefore well suited for updating the initial estimates of the market requirements with information obtained, for example, from additional market studies. Based on the updated beliefs, the optimal managerial actions can be adjusted and the future success of the product be increased. In the next section, we will therefore provide a brief overview of the Bayesian updating idea and derive a mechanism that allows to update the mean as well as the variance of the market requirement distribution. Afterwards, this information updating mechanism is integrated into a real options framework. The derived general valuation model allows to determine the project value in expectation of a later updating possibility as well as the corresponding value of the additional information.

3.2 Bayesian Updating

3.2.1 Basic Concepts of Bayesian Analysis and Inference

The basic idea of Bayesian decision making is to combine prior estimates with other relevant information about the decision problem to a posterior distribution of the state nature from which all inferences are made from (see Section 2.2.2). Before we apply this concept to our decision problem, we will briefly introduce some common notation in Bayesian analysis which will be relevant for the subsequent formulation of our model.

The unknown quantity or state of nature which affects the decision process is generally denoted as θ. It represents the unknown moments of a distribution, like the mean and/or the variance in case of a normal distribution (cf. Berger 1985; Carlin and Louis 2000). The outcome of a market study that is performed to obtain information about θ will be denoted $Z = (D_1, D_2, \ldots, D_n)$, where the D_i represent independent observations (of market requirements) from a common distribution. We will further denote a particular realization of D_i as d_i and a particular realization of Z as z. The prior information about θ is generally stated in terms of a probability distribution on θ. $\pi(\theta)$ denotes the prior density of θ, while $\pi(\theta|z)$ represents the posterior distribution of θ given the sample information (or signal) z, from which all decisions and inferences are being made.

The posterior distribution is given by the following formula (providing $m(d) \neq 0$), which is also commonly known as *Bayes' theorem*:

$$
\begin{aligned}
\pi(\theta|d) &= \frac{h(d,\theta)}{m(d)} \\
&= \frac{h(d,\theta)}{\int\limits_{-\infty}^{+\infty} h(d,\theta)d\theta} \\
&= \frac{f(d|\theta)\pi(\theta)}{\int\limits_{-\infty}^{+\infty} f(d|\theta)\pi(\theta)d\theta},
\end{aligned}
\tag{3.9}
$$

where $h(d,\theta)$ is the joint density of D and θ, $m(d)$ the marginal (unconditional) density of D, and $f(d|\theta)$ the likelihood function of D given θ.

As we will see in the subsequent model formulation, it is often convenient to re-express Bayes' theorem in the simple proportionality form

$$\pi(\theta|d) \propto f(d|\theta)\pi(\theta), \tag{3.10}$$

since the right-hand side contains all information required to reconstruct the normalizing constant,

$$\frac{1}{m(d)} = \left[\int_{-\infty}^{+\infty} f(d|\theta)\pi(\theta)d\theta \right]^{-1}, \tag{3.11}$$

when explicitly needed.

A helpful concept for calculating the posterior distribution is the one of a *sufficient statistic*. It allows to simplify the problem when the sample information consists of many independent observations. Instead of dealing with the extensive data, the decision maker can calculate a statistic and use it as a summary of the relevant information for the analysis. Such a fully informative summary of a data set can be applied if it satisfies the following definition (cf. Berger 1985, p. 35 f.):

Definition 3.1. *Let D be a random variable whose distribution depends on the unknown parameter θ, but is otherwise known. A function T of D is said to be a sufficient statistic for θ if the conditional distribution of D, given T(D) = t, is independent of θ.*

In other words, the sufficient statistic T is a function, e.g., $g(t|\theta)$, of the data that summarizes the entire sample information concerning the unknown parameter θ and hence, reduces the dimensionality of the problem. Thus, if a sufficient statistic for θ can be found, it is much easier to determine the posterior distribution over it than dealing directly with the entire data set $z = d_1, \ldots, d_n$ (cf. Berger 1985, p. 127), i.e.:

$$\pi(\theta|z) = \pi(\theta|t) = \frac{\pi(\theta)g(t|\theta)}{m(t)}. \tag{3.12}$$

The unknown quantity in our model is the unknown parameter of the market performance requirement distribution at time $t = 0$. Since the true requirements of the customers will only be known when the project has been launched, the company has to make its investment decision and start the development process based on statistical knowledge from initial market studies and/or past project experience. This information can then later be updated in the described Bayesian manner.

To do so, we have to make some assumptions. Firstly, we have to specify the model of the market performance requirements given the unknown parameter θ in form of a probability distribution $f(d|\theta)$. As stated earlier, we will assume that the requirements of the customers are normally distributed.

In this case, the unknown parameter θ can either be one-dimensional, if it represents the unknown mean or variance quantity, or two-dimensional (in this case it would be a vector, i.e., $\boldsymbol{\theta}$), if both the mean and the variance of the likelihood function are unknown.

Secondly, we have to determine the prior distribution of the random quantity $\boldsymbol{\theta}$, i.e., $\pi(\boldsymbol{\theta})$. As we will see later, there are some classes of distributions that are better suited for calculating the posterior distribution than others. Especially, if the prior belongs to the same distributional family as the likelihood $f(d|\boldsymbol{\theta})$ – is so-called *conjugate* to it, – the posterior distribution can be calculated much easier. In most other cases, the posterior density has to be determined numerically with high computational effort. Although we will use the normal conjugate relationship for our Bayesian updating model, the presented approach below can be applied to any other distributional family as well. In either case, initial studies, past project experience, or expert opinions, for example, can be used to determine the parameters of the selected prior density. It thus contains the best knowledge about the unknown quantity at that time.

The sample information for the update of the prior beliefs can be data, for example, which is obtained from additional market research studies conducted at a certain point in time after the start of the development activities. If primarily the requirements of potential customers in the target market are surveyed, the sample data is drawn from the same universe and thus, from the same distribution. On the other hand, one does not necessarily have to rely solely on market requirement data of the development project under consideration. The update can also be based on data from related projects, e.g., information obtained from the market introduction of similar products or the performance of competitors' products. This is a common approach in Bayesian updating models for supply chain management problems (cf. e.g., Fisher and Raman 1996; Eppen and Iyer 1997). In this case, however, the variance of the likelihood function reflects both the uncertainty about the unknown parameter(s) θ and the uncertainty stemming from the use of data which is obtained from similar but not the same product to update the market requirements during the development process (Iyer and Bergen 1997).

The subsequent section presents a systematic derivation of the Bayesian updating formulations needed for our model. If not stated otherwise, they follow along the reasoning of DeGroot (2004) and Berger (1985). For a more general and comprehensive discussion of Bayesian theory and analysis, the interested reader is referred to these two key references. In our model, we will consider three levels of (prior) market performance requirement uncertainty: Firstly, uncertainty about the mean, secondly, uncertainty about the variance, and finally uncertainty about both moments, the mean and the variance of the market requirement distribution (see Fig. 3.5). We will

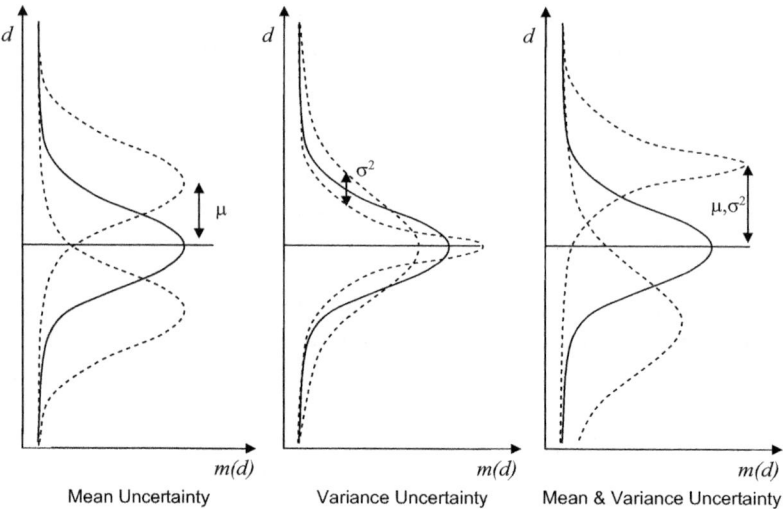

Fig. 3.5. Modeled market requirement uncertainty

always start with the prior distribution of market requirement uncertainty which is, e.g., in case of mean uncertainty, defined by the prior mean and the variance, composed of the performance requirement variance between the different customers and the uncertainty of the true mean. Afterwards, the posterior distributions of the market requirement distribution given the sample information is derived. In addition, we also derive the corresponding posterior distributions for multiple performance parameters in order to ensure generality of our model.[32]

3.2.2 Mean Update

3.2.2.1 General Properties

The first case that we will consider is the one where the firm has uncertainty about the mean, while the variance of the market performance requirement distribution is assumed to be known. This frequently assumed simplification (cf. e.g., Berger 1985; Iyer and Bergen 1997; Miller and Park 2005) is not always realistic since the firm will generally not know a priori the spread of the performance requirements among the different customers in the target market with certainty. However, if one is mainly concerned with making

---

[32]A summary of the in the following applied statistical distributions is provided in Appendix A.

inferences about the mean[33], the assumption that the variance in the market is equal to the sample variance seems to be reasonable. A sample size of over 30 is generally considered sufficient for this assumption (Antelman 1997, p. 313).

Limiting our nescience solely to the mean makes the Bayesian analysis simpler as we have to deal solely with one parameter. Despite this constraint, we will still have two levels of market requirement uncertainty in our model which are specified in the subsequent proposition.

Proposition 3.1. *Suppose the performance requirements of the customers in the target market follow a normal distribution with unknown mean θ ($\theta \in \Re$) and known variance σ^2 ($\sigma > 0$). Suppose further that the prior distribution of θ, i.e., $\pi(\theta)$, is $N(\mu, \xi^2)$ distributed.*

Then our model of market performance requirement at time 0 is a normal distribution with mean μ and variance $\sigma^2 + \xi^2$, i.e.,

$$m(d) = N(d|\mu, \sigma^2 + \xi^2). \tag{3.13}$$

Proof. The predictive belief distribution of the market performance requirement d is defined as

$$m(d) = \int_{-\infty}^{+\infty} \pi(\theta) f(d|\theta) d\theta, \tag{3.14}$$

where

$$\pi(\theta) \propto \frac{1}{\xi} \exp\left[-\frac{1}{2} \frac{(\theta - \mu)^2}{\xi^2}\right] \tag{3.15}$$

and

$$f(d|\theta) \propto \frac{1}{\sigma} \exp\left[-\frac{1}{2} \frac{(d - \theta)^2}{\sigma^2}\right] \tag{3.16}$$

if we omit all constant factors for ease of representation. Hence,

$$\pi(\theta) f(d|\theta) \propto \frac{1}{\sigma\xi} \exp\left[-\frac{1}{2}\left(\frac{(\theta - \mu)^2}{\xi^2} + \frac{(d - \theta)^2}{\sigma^2}\right)\right]. \tag{3.17}$$

It is easily verified that

$$\sigma^2(\theta - \mu)^2 + \xi^2(d - \theta)^2 = (\sigma^2 + \xi^2)\left(\theta - \frac{\sigma^2\mu + \xi^2 d}{\sigma^2 + \xi^2}\right)^2 + \frac{\sigma^2\xi^2(d - \mu)^2}{\sigma^2 + \xi^2}. \tag{3.18}$$

---

[33]This is generally the case in the Quick Response models described in Section 2.2.2.2 that primarily try to update the mean of the demand distribution of a certain product.

Hence, $m(d)$ satisfies the following relation:

$$m(d) \propto \int_{-\infty}^{+\infty} \frac{1}{\sigma\xi} \exp\left[-\frac{\sigma^2+\xi^2}{2\sigma^2\xi^2}\left(\theta - \frac{\sigma^2\mu+\xi^2 d}{\sigma^2+\xi^2}\right)^2 - \frac{(d-\mu)^2}{2(\sigma^2+\xi^2)}\right] d\theta$$

$$\propto \frac{1}{(\sigma^2+\xi^2)^{1/2}} \exp\left[-\frac{(d-\mu)^2}{2(\sigma^2+\xi^2)}\right], \tag{3.19}$$

which can be clearly recognized as belonging to a $N(\mu, \sigma^2+\xi^2)$ distribution.

Thus, the first source of uncertainty in our model captures the uncertainty about the unknown mean θ at time 0, i.e., $\pi(\theta)$, which is $N(\mu, \xi^2)$ distributed. The other one concerns the uncertainty resulting from the inhomogeneity of the potential buyers. In other words, even if we knew the true mean of the performance requirements in the market, the performance requirements varies among the different customers, i.e., $f(d|\theta) = N(d|\theta, \sigma^2)$. Hence, we obtain the total uncertainty in our model as $Var(d) = Var(\theta) + E(\sigma^2) = \xi^2 + \sigma^2$.

These two levels of uncertainty are represented by the just derived marginal distribution of market performance requirement at time 0, $m(d)$. This distribution is also sometimes called prior predictive in the Bayesian analysis literature since it summarizes our information concerning the (unobserved) market performance requirement d prior to the sample information. We will use both terms interchangeably.

In order to determine the posterior predictive market performance requirement distribution, we need to determine the posterior distribution of the unknown mean θ first. Since we will hardly update our beliefs based on a single observation, we have to deal with a sample of n independent observations $z = d_1, \ldots, d_n$. Generally, one would obtain the likelihood function $f(z|\theta)$ as $\prod_{i=1}^n f(d_i|\theta)$. From Definition 3.1 we know, however, that $T(z) = \bar{d} = \sum_{i=1}^n \frac{1}{n} d_i$ is sufficient for θ, so that we can use $f(\bar{d}|\theta) = N(\bar{d}|\theta, \sigma^2/n)$ instead for our analysis. The following proposition defines the posterior distribution of θ:

Proposition 3.2. *Suppose the performance requirements of the customers in the target market follow a normal distribution with unknown mean θ ($\theta \in \Re$) and known variance σ^2 ($\sigma > 0$) and $z = d_1, d_2, \ldots, d_n$ denotes a random sample of n independent observations of market performance requirements. If one has prior information about the mean in form of the following normal distribution $\pi(\theta) = N(\theta|\mu, \xi^2)$ ($\mu \in \Re, \xi > 0$), the posterior distribution of θ given actual data $z = d_1, d_2, \ldots, d_n$ is then given by $\pi(\theta|z) = N(\theta|\mu', \sigma'^2)$ where*

$$\mu' = \frac{\sigma^2\mu + n\xi^2\bar{d}}{\sigma^2 + n\xi^2} \tag{3.20}$$

and

$$\sigma'^2 = \frac{\sigma^2 \xi^2}{\sigma^2 + n\xi^2}.$$ (3.21)

Proof. It follows from Bayes' theorem that $\pi(\theta|z) \propto f(d_1, \dots, d_n|\theta)\pi(\theta)$. The likelihood function satisfies the following relation:

$$f(d_1, \dots, d_n|\theta) \propto \frac{1}{\sigma} \exp\left[-\frac{1}{2\sigma^2} \sum_{i=1}^{n}(d_i - \theta)^2\right].$$ (3.22)

We can further show that

$$\sum_{i=1}^{n}(d_i - \theta)^2 = \sum_{i=1}^{n}(d_i - \bar{d})^2 + 2\sum_{i=1}^{n}(d_i - \bar{d})(\bar{d} - \theta) + \sum_{i=1}^{n}(\bar{d} - \theta)^2$$

$$= n(\bar{d} - \theta)^2 + \sum_{i=1}^{n}(d_i - \bar{d})^2$$ (3.23)

since $\sum_{i=1}^{n}(d_i - \bar{d}) = 0$.

With Eq. 3.23 we obtain – if we only look at factors that involve θ [34] – for the likelihood function

$$f(d_1, \dots, d_n|\theta) \propto \exp\left[-\frac{n}{2\sigma^2}(\theta - \bar{d})^2\right]$$ (3.24)

and for the prior distribution of θ

$$\pi(\theta) \propto \exp\left[-\frac{1}{2\xi^2}(\theta - \mu)^2\right].$$ (3.25)

Hence,

$$\pi(\theta|z) \propto f(d_1, \dots, d_n|\theta)\pi(\theta)$$

$$\propto \exp\left[-\frac{n}{2\sigma^2}(\theta - \bar{d})^2 - \frac{1}{2\xi^2}(\theta - \mu)^2\right].$$ (3.26)

Under consideration of Eq. 3.18, we obtain for the posterior distribution of θ given actual data z

---

[34]This is sufficient as only these factors determine the distribution $\pi(\theta|z)$ belongs to (cf. Berger 1985, p. 130 f.); all other factors can be included in the proportionality factor which is omitted here for ease of representation.

$$\pi(\theta|z) \propto \exp\left[-\frac{1}{2}\frac{\sigma^2 + n\xi^2}{\sigma^2\xi^2}\left(\theta - \frac{\sigma^2\mu + n\xi^2\bar{d}}{\sigma^2 + n\xi^2}\right)^2\right]$$

$$\propto \exp\left[-\frac{1}{2\sigma'^2}(\theta - \mu')^2\right]. \qquad (3.27)$$

Comparing this relation with the definition of a normal distribution, we have $\pi(\theta|z) = N(\theta|\mu', \sigma'^2)$.

Defining $B = \frac{\sigma^2/n}{\sigma^2/n + \xi^2}$, we can rewrite the posterior mean as

$$\mu' = \frac{\sigma^2/n}{(\sigma^2/n) + \xi^2}\mu + \frac{\xi^2}{(\sigma^2/n) + \xi^2}\bar{d} = B\mu + (1-B)\bar{d} \qquad (3.28)$$

and the posterior variance as

$$\sigma'^2 = (1-B)\sigma^2/n. \qquad (3.29)$$

This representation shows that the posterior mean is a weighted average of the prior mean μ and the observed data value, which is summarized by the sample mean $\bar{d}$. Since the weights are hereby inversely proportional to the variances, it follows that the larger the sample size n and the smaller the variance of each observation σ^2 relative to ξ^2 – which corresponds to a vague prior information – the closer will be the posterior mean to the observed data $\bar{d}$. The posterior variance, on the other hand, decreases as the number of observations increases. Thus, the posterior mean combines the subjective prior information with the observed data and provides with these two sources of information an increased precision of θ.

Based on these considerations, we obtain the following posterior predictive market requirement distribution for our model:

Corollary 3.1. *Suppose the uncertainties are given as specified in Proposition 3.2. After the update with the actual data* $z = d_1, d_2, \ldots, d_n$ *obtained from the follow-up study, the model of market performance requirement at time* τ *is then a normal distribution with mean* μ' *and variance* $\sigma^2/n + \sigma'^2$, *i.e.,* $m(d|z) = N(d|\mu', \sigma^2/n + \sigma'^2)$.

Proof. It follows in direct analogy to the proof of Proposition 3.1 under consideration of Proposition 3.2. We know from Proposition 3.1 that the prior marginal distribution is given as $m(d) = N(d|\mu, \sigma^2 + \xi^2)$ if the unknown mean is normally distributed, i.e., $\pi(\theta) = N(\theta|\mu, \xi^2)$, while the variance σ^2 is known. The update of the initial estimates about θ with the additional sample z results a posterior distribution of $\pi(\theta|z) = N(\theta|\mu', \sigma'^2)$, i.e., the mean changes from μ to μ' and the variance from ξ^2 to σ'^2 (Proposition 3.2).

We further know that the posterior marginal distribution of the market performance requirement d is defined as

$$m(d|z) = \int\limits_{-\infty}^{+\infty} \pi(\theta|z)f(d|\theta)d\theta, \tag{3.30}$$

where

$$\pi(\theta|z) \propto \exp\left[-\frac{1}{2\sigma'^2}(\theta - \mu')^2\right] \tag{3.31}$$

and

$$f(d_1,\ldots,d_n|\theta) \propto \exp\left[-\frac{n}{2\sigma^2}(\theta - \bar{d})^2\right] \tag{3.32}$$

if we only look – for ease of representation – at factors that involve θ.

Under consideration of Eq. 3.18 and the results of Proposition 3.2, the following relation holds for $m(d|z)$:

$$m(d|z) \propto \int\limits_{-\infty}^{+\infty} \frac{1}{(\sigma/n)\sigma'} \exp\left[-\frac{\sigma^2 + n\sigma'^2}{2\sigma^2\sigma'^2}\left(\theta - \frac{\sigma^2\mu + n\sigma'^2 d}{\sigma^2 + n\sigma'^2}\right)^2\right.$$
$$\left. - \frac{(d - \mu')^2}{2(\sigma^2/n + \sigma'^2)}\right] d\theta$$
$$\propto \frac{1}{(\sigma^2/n + \sigma'^2)^{1/2}} \exp\left[-\frac{(d - \mu')^2}{2(\sigma^2/n + \sigma'^2)}\right]. \tag{3.33}$$

Comparing this result with the definition of a normal distribution, we have $m(d|z) = N(\mu', \sigma^2/n + \sigma'^2)$.

Thus, the posterior variance of the predictive believe distribution for a single observable consists, similar to the prior marginal distribution, of two elements (cf. e.g., Iyer and Bergen 1997): $Var(d|z) = Var(\theta|z) + E(\sigma^2|z) = \sigma'^2 + \sigma^2/n$. It summarizes our information concerning the market performance requirement d after having observed the data from the follow-up study z.

3.2.2.2 Multidimensional Case

In the derivation of the market performance requirement update, we have solely been concerned with a single product attribute so far, i.e., a single dimensional performance parameter. If we consider k performance parameters in the valuation of a new product development project as presented in Section 3.1.5, we need a multivariate Bayesian estimation of the unknown mean vector $\theta = (\theta_1, \theta_2, \ldots, \theta_k)'$.[35] The approach is similar to the univariate case and thus, straightforward.

Once again, we assume that the performance requirements in the target market are normally distributed. All samples are therefore taken from a nonsingular, k-dimensional multivariate normal distribution, where any observation X will be a k-dimensional, continuous random vector whose value x will be an element of $\Re^k$. The mean vector θ will also be k-dimensional and the variance matrix Σ will be a $k \times k$ symmetric, positive-definite matrix.

Proposition 3.3. *Suppose the performance requirements of the customers in the target market follow a normal distribution with unknown mean vector θ ($\theta \in \Re^k$) and known covariance matrix Σ (Σ is a $k \times k$ symmetric, positive-definite matrix) and $z = d_1, d_2, \ldots, d_n$ denotes a random sample of n independent (k-dimensional) observations of market performance requirements. If one has prior information about the mean vector in form of the following multivariate normal distribution $\pi(\theta) = N_k(\theta|\mu, A)$ ($\mu \in \Re^k$, A is a $k \times k$ symmetric, positive-definite matrix), the posterior distribution of the mean vector θ given actual data z is then given by $\pi(\theta|z) = N_k(\theta|\mu^p, \Sigma^p)$ where*

$$\mu^p = (\Sigma + nA)^{-1}(\Sigma\mu + nA\bar{z}) \tag{3.34}$$

and

$$\Sigma^p = (A^{-1} + n\Sigma^{-1})^{-1}. \tag{3.35}$$

Proof. The proof follows in analogy to the proof of the univariate case as provided for Proposition 3.2. The likelihood function satisfies the following relation:

$$f_n(d_1, d_2, \ldots, d_n|\theta) \propto \exp\left[-\frac{1}{2}\sum_{i=1}^{n}(d_i - \theta)'\Sigma^{-1}(d_i - \theta)\right]. \tag{3.36}$$

Since

$$\sum_{i=1}^{n}(d_i - \theta)'\Sigma^{-1}(d_i - \theta)$$

$$= \sum_{i=1}^{n}(d_i - \bar{d})'\Sigma^{-1}(d_i - \bar{d}) + n(\theta - \bar{d})'\Sigma^{-1}(\theta - \bar{d}), \tag{3.37}$$

we can rewrite relation Eq. 3.36 – if we only look at factors that involve θ – as

---

[35]Note that θ' denotes in the multidimensional case a transposed vector. Posterior values will therefore be denoted by superscript p, e.g., μ^p.

$$f_n(d_1, d_2, \ldots, d_n | \theta) \propto \exp\left[-\frac{1}{2}(\theta - \bar{d})' n \Sigma^{-1}(\theta - \bar{d})\right]. \tag{3.38}$$

The prior distribution of the mean vector satisfies the following relation:

$$\pi(\theta) \propto \exp\left[-\frac{1}{2}(\theta - \mu)' A^{-1}(\theta - \mu)\right]. \tag{3.39}$$

Based on Bayes' theorem that $\pi(\theta | z) \propto f_k(d_1, d_2, \ldots, d_n | \theta) \pi(\theta)$, we obtain the following relation for the posterior distribution of the mean vector:

$$\pi(\theta | z) \propto \exp\left[-\frac{1}{2}(\theta - \mu)' A^{-1}(\theta - \mu) - \frac{1}{2}(\theta - \bar{d})' n \Sigma^{-1}(\theta - \bar{d})\right]. \tag{3.40}$$

However, in analogy to Eq. 3.18, it can be shown that

$$(\theta - \mu)' A^{-1}(\theta - \mu) + (\theta - \bar{d})' n \Sigma^{-1}(\theta - \bar{d})$$
$$= (\theta - \mu^p)'(A^{-1} + n\Sigma^{-1})(\theta - \mu^p) + (\theta - \bar{d})'(A^{-1} + n\Sigma^{-1})(\theta - \bar{d}), \tag{3.41}$$

with

$$\mu^p = (\Sigma + nA)^{-1}(\Sigma\mu + nA\bar{d})$$
$$= (A^{-1} + n\Sigma^{-1})^{-1}(A^{-1}\mu + n\Sigma^{-1}\bar{d}). \tag{3.42}$$

Thus, Eq. 3.40 can be rewritten as

$$\pi(\theta | z) \propto \exp\left[-\frac{1}{2}(\theta - \mu^p)'(A^{-1} + n\Sigma^{-1})(\theta - \mu^p)\right], \tag{3.43}$$

which corresponds to the density of a multivariate normal distribution with the mean vector and covariance matrix as specified above.

3.2.3 Variance Update

3.2.3.1 General Properties

In the previous section, we considered the case that the firm has uncertainty about the mean only, while the variance of the market performance requirement distribution is known. In what follows, we shall consider the opposite case where the company has uncertainty about the true variance while the value of the mean can be specified. The market performance requirement uncertainty in our model can thus be specified as follows:

Proposition 3.4. *Suppose the performance requirements of the customers in the target market follow a normal distribution with known mean* μ *(*$\mu \in \Re$*) and un-*

known variance σ^2 ($\sigma > 0$). Suppose further that the prior distribution $\pi(\sigma^2)$ is inverse gamma distributed, i.e., $IG(\alpha, \beta)$ ($\alpha > 0, \beta > 0$).[36]
    Then our model of market performance requirement at time 0 is t distributed, i.e.,

$$m(d) = St(d|\mu, (\alpha\beta)^{-1}, 2\alpha). \tag{3.44}$$

Proof. The predictive belief distribution of the market performance requirement d is defined as

$$m(d) = \int_0^{+\infty} \pi(\sigma^2) f(d|\sigma^2) d\sigma^2, \tag{3.45}$$

where

$$\pi(\sigma^2) \propto \frac{1}{(\sigma^2)^{\alpha+1}} \exp\left[-\frac{1}{\beta\sigma^2}\right] \tag{3.46}$$

and

$$f(d|\sigma^2) \propto \frac{1}{\sigma} \exp\left[-\frac{1}{2\sigma^2}(d - \mu)^2\right] \tag{3.47}$$

if we omit all constant factors for ease of representation. Hence,

$$\pi(\sigma^2) f(d|\sigma^2) \propto \frac{1}{(\sigma^2)^{\alpha+1}} \exp\left[-\frac{1}{\beta\sigma^2}\right] \frac{1}{\sigma} \exp\left[-\frac{1}{2\sigma^2}(d - \mu)^2\right] \tag{3.48}$$

and we obtain the following relation for $m(d)$:

$$m(d) \propto \int_0^{+\infty} \frac{1}{(\sigma^2)^{\alpha+3/2}} \exp\left[-\frac{1}{\beta\sigma^2} - \frac{1}{2\sigma^2}(d - \mu)^2\right] d\sigma^2$$

$$\propto \left[1 + \frac{\beta\alpha(d - \mu)^2}{2\alpha}\right]^{-\frac{2\alpha+1}{2}}, \tag{3.49}$$

which corresponds to the density function of a $St(\mu, (\alpha\beta)^{-1}, 2\alpha)$ distribution.

   The determination of a posterior density of θ is generally rather difficult since most integrals can only be evaluated numerically. In Bayesian statistics one frequently chooses therefore, as mentioned earlier, a prior distribution for which the posterior distribution can easily be determined. These so-called conjugate priors are selected in such a way that they are *conjugate* to the likelihood $f(z|\theta)$. This approach has the property that the posterior dis-

---

[36]This prior distribution is chosen as it is conjugate to the selected likelihood function. See the subsequent discussion for details.

tribution belongs to the same distributional family as the prior distribution regardless of the sample size n and the observed sample data.

We will apply this property here. If we thus look at the density function of the normal distribution as a function of the variance σ^2, we can detect a proportionality to $(\sigma^2)^{-a} \exp(-b/\sigma^2)$ (a and b representing constant terms), which resembles in fact the density function of the inverse gamma distribution (see Appendix A.1.3). Based on this consideration, we can determine the posterior distribution of the unknown variance as follows:

Proposition 3.5. *Suppose the performance requirements of the customers in the target market follow a normal distribution with known mean μ ($\mu \in \Re$) and unknown variance σ^2 ($\sigma > 0$) and $z = d_1, d_2, \ldots, d_n$ denotes a random sample of n independent observations of market performance requirements. If the prior distribution of the variance is an inverse gamma distribution, i.e., $\pi(\sigma^2) = IG(\sigma^2|\alpha, \beta)$ ($\alpha > 0, \beta > 0$), the posterior distribution of σ^2 given actual data $z = d_1, d_2, \ldots, d_n$ is then given by $\pi(\sigma^2|z) = IG(\sigma^2|\alpha', \beta')$, where*

$$\alpha' = \alpha + \frac{n}{2} \tag{3.50}$$

and

$$\beta' = \left[\frac{1}{\beta} + \frac{1}{2} \sum_{i=1}^{n} (d_i - \mu)^2 \right]^{-1}. \tag{3.51}$$

Proof. It follows from Bayes' theorem that $\pi(\sigma^2|z) \propto f(d_1, \ldots, d_n|\sigma^2)\pi(\sigma^2)$. The likelihood function satisfies the following relation:

$$f(d_1, \ldots, d_n|\sigma^2) \propto \frac{1}{(\sigma^2)^{n/2}} \exp\left[-\frac{1}{2\sigma^2} \sum_{i=1}^{n} (d_i - \mu)^2 \right] \tag{3.52}$$

and the prior distribution of σ^2 is given by

$$\pi(\sigma^2) \propto \frac{1}{(\sigma^2)^{\alpha+1}} \exp\left[-\frac{1}{\beta\sigma^2} \right]. \tag{3.53}$$

Hence, we obtain for the posterior distribution of σ^2 given actual data z

$$\pi(\sigma^2|z) \propto \frac{1}{(\sigma^2)^{\alpha+n/2+1}} \exp\left[-\frac{1}{\beta\sigma^2} - \frac{\sum_{i=1}^{n}(d_i-\mu)^2}{2\sigma^2} \right]$$

$$\propto \frac{1}{(\sigma^2)^{\alpha+n/2+1}} \exp\left[-\frac{1}{\sigma^2}\left(\frac{1}{\beta} + \frac{1}{2}\sum_{i=1}^{n}(d_i-\mu)^2 \right) \right]$$

$$\propto \frac{1}{(\sigma^2)^{\alpha'+1}} \exp\left[-\frac{1}{\sigma^2\beta'} \right]. \tag{3.54}$$

Comparing this relation with the definition of the inverse gamma density function (Appendix A.1.3), we can see that $\pi(\sigma^2|z)$ is $IG(\alpha',\beta')$ distributed.

Our model of market requirement uncertainty given the additional information can thus be determined as follows:

Corollary 3.2. *Suppose the uncertainties are given as defined in Proposition 3.5. After the update with the actual data* $z = d_1, d_2, \ldots, d_n$ *obtained from the follow-up study, the model of market performance requirement at time* τ *is then t distributed, i.e.,*

$$m(d|z) = St\left(d|\mu, (\alpha'\beta')^{-1}, 2\alpha + n\right). \tag{3.55}$$

Proof. It follows in direct analogy the proof of Proposition 3.4 under consideration of Proposition 3.5. We know from Proposition 3.4 that the prior marginal distribution is given as $m(d) = St(d|\mu, (\alpha\beta)^{-1}, 2\alpha)$ if the unknown variance is inverse gamma distributed, i.e., $\pi(\sigma^2) = IG(\sigma^2|\alpha, \beta)$, while the mean μ is known. The update of the initial estimates about σ^2 with the additional sample z results a posterior distribution of $\pi(\sigma^2|z) = IG(\sigma^2|\alpha', \beta')$ (Proposition 3.5).

We further know that the posterior marginal distribution of the market performance requirement d is defined as

$$m(d|z) = \int_0^{+\infty} \pi(\sigma^2|z) f(d|\sigma^2) d\sigma^2, \tag{3.56}$$

where

$$\pi(\sigma^2|z) \propto \frac{1}{(\sigma^2)^{\alpha'+1}} \exp\left[-\frac{1}{\sigma^2\beta'}\right] \tag{3.57}$$

and

$$f(d_1, \ldots, d_n|\sigma^2) \propto \frac{1}{(\sigma^2)^{n/2}} \exp\left[-\frac{1}{2\sigma^2} \sum_{i=1}^n (d_i - \mu)^2\right] \tag{3.58}$$

if we omit all constant factors for ease of representation.

Under consideration of Proposition 3.5, the following relation holds for

$$m(d|z) \propto \int_0^{+\infty} \frac{1}{(\sigma^2)^{\alpha'+3/2}} \exp\left[-\frac{1}{\beta'\sigma^2} - \frac{1}{2\sigma^2}(d - \mu)^2\right] d\sigma^2$$

$$\propto \left[1 + \frac{\beta'\alpha'(d - \mu)^2}{2\alpha'}\right]^{-\frac{2\alpha'+1}{2}}. \tag{3.59}$$

Thus, the posterior marginal distribution is t distributed with mean μ, variance $(\alpha'\beta')^{-1}$, and $(2\alpha + n)$ degrees of freedom. The n additional degrees of freedom hereby result from the sample of size n, i.e., $z = d_1, d_2, \ldots, d_n$.

Simple algebra also shows here again that the uncertainty of the posterior predictive believe distribution for a single observable complies with the expected value of the posterior distribution of σ^2, i.e., $\pi(\sigma^2|z)$, namely $Var(d|z) = Var(\mu|z) + E(\sigma^2|z) = [\beta'(\alpha' - 1)]^{-1}$.

3.2.3.2 Multidimensional Case

The generalization of this case, where we know the mean vector but not the covariance matrix of the multivariate normally distributed performance requirements, directly follows from Proposition 3.5. In the subsequent proposition, the reciprocal of the Wishart distribution (the inverse Wishart) represents the generalization of the inverted gamma distribution (cf. Carlin and Louis 2000, p. 328). To avoid the rather awkward inverse Wishart distribution, one could alternatively derive the results for the precision[37] rather than for the covariance matrix, which is generally more convenient. For reasons of consistency, however, we will stay with the former distribution.

Proposition 3.6. *Suppose the performance requirements of the customers in the target market follow a normal distribution with mean vector μ and unknown covariance matrix Ψ and $z = d_1, d_2, \ldots, d_n$ denotes a random sample of n independent (k-dimensional) observations of market performance requirements. If one has prior information about the covariance matrix Ψ, i.e., $\pi(\psi)$, in form of an inverse Wishart distribution with covariance matrix Σ and α degrees of freedom (Σ is a $k \times k$ symmetric, positive-definite matrix, $\alpha > k - 1$), the posterior distribution of ψ given actual data z, i.e., $\pi(\psi|z)$, is then an inverse Wishart distribution with covariance matrix Σ^p and $\alpha + n$ degrees of freedom, where*

$$\Sigma^p = \left(\Sigma^{-1} + \sum_{i=1}^{n}(d_i - \mu)(d_i - \mu)' \right)^{-1}. \tag{3.60}$$

Proof. The proof follows in analogy to the proof of the univariate case as provided for Proposition 3.5. The multivariate normally distributed likelihood function satisfies the following relation:

$$f_n(d_1, d_2, \ldots, d_n|\psi) \propto |\psi|^{-n/2} \exp\left[-\frac{1}{2} \sum_{i=1}^{n}(d_i - \mu)'\psi^{-1}(d_i - \mu) \right], \tag{3.61}$$

---

[37] The reciprocal of the variance is called precision.

where $|\cdot|$ denotes the determinant of a matrix. Since the exponent of the e function in relation Eq. 3.61 is a real number and hence, a 1x1 matrix, the following holds:

$$\sum_{i=1}^{n}(d_i - \mu)'\psi^{-1}(d_i - \mu) = tr\left[\sum_{i=1}^{n}(d_i - \mu)'\psi^{-1}(d_i - \mu)\right]$$

$$= tr\left\{\left[\sum_{i=1}^{n}(d_i - \mu)'(d_i - \mu)\right]\psi^{-1}\right\}, \quad (3.62)$$

where $tr(\cdot)$ denotes the trace of a matrix argument.[38]

The prior distribution of the covariance matrix ψ satisfies the following relation:

$$\pi(\psi) \propto |\psi|^{-(\alpha-k-1)/2}\exp\left[-\frac{1}{2}tr(\Sigma^{-1}\psi^{-1})\right]. \quad (3.63)$$

Based on Bayes' theorem that $\pi(\psi|z) \propto f_k(d_1, d_2, \ldots, d_n|\psi)\pi(\psi)$, we obtain the following relation for the posterior distribution of the mean vector:

$$\pi(\psi|z)$$

$$\propto |\psi|^{-(\alpha+n-k-1)/2}\exp\left[-\frac{1}{2}tr\left[\left(\sum_{i=1}^{n}(d_i - \mu)'(d_i - \mu)\right)\psi^{-1} + (\Sigma^{-1}\psi^{-1})\right]\right]$$

$$\propto |\psi|^{-(\alpha+n-k-1)/2}\exp\left[-\frac{1}{2}tr\left[\Sigma^{-1} + \sum_{i=1}^{n}(d_i - \mu)'(d_i - \mu)\right]\psi^{-1}\right], \quad (3.64)$$

which corresponds to the density of a inverse Wishart distribution with the mean vector and covariance matrix as specified above.

So far we have considered two special cases where the firm always has uncertainty about just one parameter of the market requirement distribution (either the mean or the variance), while the other one is assumed to be known. In the next section, we will relax this limitation and provide the general model that captures uncertainty about both the mean and the variance.

3.2.4 Mean and Variance Update

3.2.4.1 General Properties

The most common case is the one where the company has some uncertainty about both the true mean and the true variance of the market performance

---

[38] For a definition, see Bronstein et al. (1999, p. 251), for example.

requirement distribution prior to the start of a development project. This implies that management can neither specify the mean nor the variance with high precision. However, the company will generally have some prior information about both moments obtained, for example, from initial market studies, previous projects, or expert opinions. This case can therefore be regarded as a combination of the two previous ones with the following sources of uncertainty in our market performance requirement model, specified in the subsequent proposition.

Proposition 3.7. *Suppose the performance requirements of the customers in the target market follow a normal distribution with unknown mean θ ($\theta \in \Re$) and unknown variance σ^2 ($\sigma > 0$). Suppose further that the joint prior distribution of the mean and the variance is $\pi(\theta, \sigma^2) = \pi_1(\theta|\sigma^2)\pi_2(\sigma^2)$, where $\pi_1(\theta|\sigma^2)$ is a $N(\mu, v\sigma^2)$ density ($\mu \in \Re, v > 0$) and $\pi_2(\sigma^2)$ is an $IG(\alpha, \beta)$ density ($\alpha > 0, \beta > 0$).*

Then our model of market performance requirement at time 0 is t distributed, i.e.,

$$m(d) = St\left(d|\mu, \frac{v+1}{\alpha\beta}, 2\alpha\right), \tag{3.65}$$

with $E(d) = \mu$ and $Var(d) = Var(\theta) + E(\sigma^2) = \frac{v+1}{(\alpha-1)\beta}$.[39]

Proof. The predictive belief distribution of the market performance requirement d is defined as (cf. Bernardo and Smith 2000, p. 378 f.)

$$m(d) = \int_{-\infty}^{+\infty}\int_{0}^{+\infty} f(d|\theta, \sigma^2)\pi(\theta, \sigma^2)d\theta d\sigma^2. \tag{3.66}$$

The likelihood function satisfies the following relation:

$$f(d|\theta, \sigma^2) \propto \frac{1}{\sigma}\exp\left[-\frac{1}{2\sigma^2}(d-\theta)^2\right]. \tag{3.67}$$

We further know that

$$\pi(\theta|\sigma^2) \propto \frac{1}{\sigma}\exp\left[-\frac{1}{2v\sigma^2}(\theta-\mu)^2\right] \tag{3.68}$$

and

$$\pi(\sigma^2) \propto \frac{1}{(\sigma^2)^{\alpha+1}}\exp\left[-\frac{1}{\beta\sigma^2}\right], \tag{3.69}$$

so that we obtain

---

[39] For the marginal density function of θ, $\pi(\theta)$, see Corollary 3.3.

$$\pi(d|\theta,\sigma^2)\pi(\theta,\sigma^2) \propto -\frac{1}{(\sigma^2)^{\alpha+2}} \exp\left[\frac{(d-\theta)^2}{2\sigma^2} + \frac{(\theta-\mu)^2}{2v\sigma^2} + \frac{1}{\beta\sigma^2}\right]$$

$$\propto -\frac{1}{(\sigma^2)^{\alpha+2}} \exp\left[\frac{1}{2v\sigma^2}\left[v(d-\theta)^2 + (\theta-\mu)^2\right] + \frac{1}{\beta\sigma^2}\right].$$
(3.70)

Under consideration of Eq. 3.18, the density function of the prior predictive believe distribution can be specified by relation

$$m(d) \propto \int_{-\infty}^{+\infty}\int_{0}^{+\infty} -\frac{1}{(\sigma^2)^{\alpha+2}} \exp\left[\frac{v+1}{2v\sigma^2}\left(\theta - \frac{vd+\mu}{v+1}\right)^2 + \frac{(d-\mu)^2}{2\sigma^2(v+1)}\right.$$

$$\left. + \frac{1}{\beta\sigma^2}\right] d\theta d\sigma^2$$

$$\propto \int_{0}^{+\infty} -\frac{1}{(\sigma^2)^{\alpha+3/2}} \exp\left[\frac{(d-\mu)^2}{2\sigma^2(v+1)} + \frac{1}{\beta\sigma^2}\right] d\sigma^2$$

$$\propto \left[\frac{(d-\mu)^2}{2\sigma^2(v+1)} + \frac{1}{\beta}\right]^{-\frac{2\alpha+1}{2}}$$

$$\propto \left[1 + \frac{1}{2\alpha}\frac{\alpha\beta(d-\mu)^2}{v+1}\right]^{-\frac{2\alpha+1}{2}},$$
(3.71)

which resembles the density function of a $St(\mu, \frac{v+1}{\alpha\beta}, 2\alpha)$ distribution. This proves the first statement.

We know by definition that

$$Var(\theta) = \frac{2\alpha v}{(2\alpha-2)\alpha\beta}$$
(3.72)

and

$$E(\sigma^2) = \frac{1}{\beta(\alpha-1)}.$$
(3.73)

Thus

$$Var(\theta) + E(\sigma^2) = \frac{v+1}{(\alpha-1)\beta}$$
(3.74)

which corresponds to the variance of the market performance requirement, $Var(d)$, and the second statement is proven.

In order to determine the posterior value of the unknown mean and variance, we will assume again the already known appropriate conjugate priors for these two unknown quantities. The next proposition specifies the corresponding posterior distributions.

Proposition 3.8. *Suppose the performance requirements of the customers in the target market follow a normal distribution with unknown mean θ ($\theta \in \Re$) and unknown variance σ^2 ($\sigma > 0$) and $z = d_1, d_2, \ldots, d_n$ denotes a random sample of n independent observations of market performance requirements. Suppose further that the joint prior distribution of the mean and the variance is*

$$\pi(\theta, \sigma^2) = \pi_1(\theta|\sigma^2)\pi_2(\sigma^2),$$

where $\pi_1(\theta|\sigma^2)$ is a $N(\mu, v\sigma^2)$ density ($\mu \in \Re, v > 0$) and $\pi_2(\sigma^2)$ is an $IG(\alpha, \beta)$ density ($\alpha > 0, \beta > 0$).

The joint posterior distribution of θ and σ^2 given actual data $z = d_1, d_2, \ldots, d_n$ is then given by

$$\pi(\theta, \sigma^2|z) = \pi_1(\theta|\sigma^2, z)\pi_2(\sigma^2|z),$$

where

$$\pi_1(\theta|\sigma^2, z) = N\left(\theta|\mu', \frac{\sigma^2}{v^{-1} + n}\right) \tag{3.75}$$

with

$$\mu' = \frac{\mu + nv\bar{d}}{nv + 1} \tag{3.76}$$

and

$$\pi_2(\sigma^2|z) = IG\left(\sigma^2|\alpha + \frac{n}{2}, \beta'\right) \tag{3.77}$$

with

$$\beta' = \left[\frac{1}{\beta} + \frac{1}{2}\sum_{i=1}^{n}(d_i - \bar{d})^2 + \frac{n(\bar{d} - \mu)^2}{2(1 + nv)}\right]^{-1}. \tag{3.78}$$

Proof. We know from Bayes' theorem that the joint posterior distribution of θ and σ^2 given data z satisfies the following relation:

$$\pi(\theta, \sigma^2|z) \propto f(d_1, \ldots, d_n|\theta, \sigma^2)\pi(\theta, \sigma^2). \tag{3.79}$$

Since the observations d_i are independent, the likelihood function can be determined as

$$f(d_1, \ldots, d_n|\theta, \sigma^2) = f(d_1|\mu, \sigma^2) \times \ldots \times f(d_n|\mu, \sigma^2)$$

and the following relation holds

$$f(d_1, \ldots, d_n|\theta, \sigma^2) \propto \frac{1}{(\sigma^2)^{n/2}} \exp\left[\frac{1}{2\sigma^2}\sum_{i=1}^{n}(d_i - \theta)^2\right]. \tag{3.80}$$

The joint prior distribution of θ and σ^2 satisfies the following relation:

$$\pi(\theta, \sigma^2) \propto \frac{1}{(\sigma^2)^{\alpha+3/2}} \exp\left[\frac{1}{2v\sigma^2}(\theta - \mu)^2 - \frac{1}{\beta\sigma^2}\right]. \tag{3.81}$$

Hence, the joint posterior distribution of θ and σ^2 given actual data $z = d_1, d_2, \ldots, d_n$ can be written as

$$\pi(\theta, \sigma^2|z) \propto \frac{1}{(\sigma^2)^{\alpha+3/2+n/2}} \exp\left[\frac{1}{2v\sigma^2}\left(v\sum_{i=1}^{n}(d_i - \theta)^2 + (\theta - \mu)^2\right) - \frac{1}{\beta\sigma^2}\right]. \tag{3.82}$$

With

$$\sum_{i=1}^{n}(d_i - \theta)^2 = n(\bar{d} - \theta)^2 + \sum_{i=1}^{n}(d_i - \bar{d})^2 ,$$

from Eq. 3.18 and

$$nv(\bar{d} - \theta)^2 + (\theta - \mu)^2 = (nv + 1)\left(\theta - \frac{nv\bar{d} + \mu}{nv + 1}\right)^2 + \frac{nv(\mu - \bar{d})^2}{nv + 1} ,$$

from Eq. 3.23, we can rewrite Eq. 3.82 as follows:

$$\pi(\theta, \sigma^2|z) \propto \frac{1}{\sigma} \exp\left[-\frac{nv + 1}{2v\sigma^2}\left(\theta - \frac{nv\bar{d} + \mu}{nv + 1}\right)^2\right] \exp\left[-\frac{1}{\sigma^2}\right]$$
$$\left(\frac{1}{2}\sum_{i=1}^{n}(d_i - \bar{d})^2 + \frac{n(\mu - \bar{d})^2}{2(nv + 1)} + \frac{1}{\beta}\right)\right] \frac{1}{(\sigma^2)^{\alpha+n/2+1}}$$
$$\propto \frac{1}{\sigma} \exp\left[-\frac{nv + 1}{2v\sigma^2}(\theta - \mu')^2\right] \exp\left[-\frac{1}{\sigma^2\beta'}\right] \frac{1}{(\sigma^2)^{\alpha+n/2+1}} \tag{3.83}$$

where μ' is defined by Eq. 3.76 and β' by Eq. 3.78.

The first two factors of the last expression are recognizable as belonging to the $N(\mu', \frac{\sigma^2}{v^{-1}+n})$ distribution of $\pi_1(\theta|\sigma^2, z)$, while the last two factors correspond to the $IG(\alpha + n/2, \beta')$ distribution of $\pi_2(\sigma^2|z)$.

As seen in Proposition 3.8, we need the prior conditional distribution of mean θ as a starting point for the update. While the determination of this conditional distribution is generally very difficult, it is much easier to estimate the key moments (i.e., mean and variance) of the corresponding marginal distribution and derive afterwards the key parameters of the conditional distribution. The prior marginal density of θ is specified below.

Corollary 3.3. *Suppose the conditional distribution of mean θ, i.e., $\pi_1(\theta|\sigma^2)$, is given as defined in Proposition 3.5. Then the marginal probability density function of θ has the form*

$$\pi(\theta) = St\left(\mu, \frac{\nu}{\alpha\beta}, 2\alpha\right). \tag{3.84}$$

Proof. For $\theta \in \Re$, the marginal density function of θ is defined by

$$\pi(\theta) = \int_0^{+\infty} \pi(\theta, \sigma^2) d\sigma^2, \tag{3.85}$$

where $\pi(\theta, \sigma^2)$ is defined as in Proposition 3.8. If we look only at the factors that involve θ, we obtain the following relation:

$$\pi(\theta) \propto \int_0^{+\infty} \frac{1}{(\sigma^2)^{\alpha+3/2}} \exp\left[\frac{1}{2\nu\sigma^2}(\theta-\mu)^2 - \frac{1}{\beta\sigma^2}\right] d\sigma^2$$

$$\propto \left[\frac{(\theta-\mu)^2}{2\sigma^2\nu} + \frac{1}{\beta}\right]^{-\frac{2\alpha+1}{2}}$$

$$\propto \left[1 + \frac{1}{2\alpha}\frac{\alpha\beta(\theta-\mu)^2}{\nu}\right]^{-\frac{2\alpha+1}{2}}. \tag{3.86}$$

This corresponds to the density function of the t distribution with the parameters given above.

The following proposition specifies the corresponding posterior distribution:

Corollary 3.4. *The posterior marginal distribution of θ is a t distribution of the following form*

$$\pi(\theta|z) = St\left(\theta|\mu', \left[(n+\frac{1}{\nu})(\alpha+\frac{n}{2})\beta'\right]^{-1}, 2\alpha+n\right). \tag{3.87}$$

Proof. Follows directly from Corollary 3.3 if we replace the parameters μ, ν, α, and β of the prior distribution $\pi(\theta)$ with their posterior values as derived in Proposition 3.5.

Based on these properties for the general case of uncertainty about the mean as well as the variance, we can derive the following property for the posterior predictive distribution of the market requirement:

Corollary 3.5. *Suppose the uncertainties are given as defined in Proposition 3.5. After the update with the actual data $z = d_1, d_2, \ldots, d_n$ obtained from the follow-up study, the model of market performance requirement at time τ is then t distributed, i.e.,*

$$m(d|z) = St\left(d|\mu', \left[\frac{(v^{-1}+n)(\alpha+\frac{n}{2})\beta'}{v^{-1}+n+1}\right]^{-1}, 2\alpha+n\right), \qquad (3.88)$$

with $E(d|z) = \mu'$ and $Var(d|z) = Var(\theta|z) + E(\sigma^2|z) = \frac{v^{-1}+n+1}{(\alpha+n/2-1)(v^{-1}+n)\beta'}$.

Proof. The proof of the first statement is straightforward. Since it follows in direct analogy the proof of Proposition 3.7 under consideration of Proposition 3.8, we will only show the analogies between the prior and posterior predictive market requirement distribution.

We know from Proposition 3.7 that the prior marginal distribution is given as $m(d) = St\left(d|\mu, \frac{v+1}{\alpha\beta}, 2\alpha\right)$ if one has estimates about the unknown mean θ and unknown variance σ^2 in form of the following prior distributions: $\pi_1(\theta|\sigma^2) = N(\theta|\mu, v\sigma^2)$ and $\pi_2(\sigma^2) = IG(\sigma^2|\alpha, \beta)$. The update of these initial estimates with the sample z results in the following two posterior distributions: $\pi_1(\theta|\sigma^2, z) = N\left(\theta|\mu', \frac{\sigma^2}{v^{-1}+n}\right)$ and $\pi_2(\sigma^2|z) = IG\left(\sigma^2|\alpha+\frac{n}{2}, \beta'\right)$. Thus, if we replace all prior parameters with the corresponding posterior values, we obtain a posterior marginal distribution with mean μ', variance $\left[\frac{(v^{-1}+n)(\alpha+\frac{n}{2})\beta'}{v^{-1}+n+1}\right]^{-1}$, and $(2\alpha+n)$ degrees of freedom.[40] This proves the first statement.

We know by definition that

$$Var(\theta|z) = \frac{2\alpha+n}{(2\alpha+n-2)(v^{-1}+n)(\alpha+n/2)\beta'} \qquad (3.89)$$

and

$$E(\sigma^2) = \frac{1}{(\alpha+n/2-1)\beta'}. \qquad (3.90)$$

Thus

$$Var(\theta) + E(\sigma^2) = \frac{v^{-1}+n+1}{(\alpha+n/2-1)(v^{-1}+n)\beta'} \qquad (3.91)$$

which corresponds the variance of the market performance requirement, $Var(d|z)$, and the second statement is proven.

3.2.4.2 Illustrative Numerical Example

Before we provide the posterior market requirement distribution for the multidimensional case, we will illustrate the just derived general Bayesian updating mechanism with the help of a numerical example.

---

[40]Pleas note that $v' = \frac{v}{1+nv}$ and that the n additional degrees of freedom result from the sample of size n, i.e., $z = d_1, d_2, \ldots, d_n$.

Table 3.1. Estimated moments of prior distribution

Unknown parameter	Moment	Value
Mean θ	$E(\theta)$	0
	$Var(\theta)$	2
Variance σ^2	$E(\sigma^2)$	2
	$Var(\sigma^2)$	5

Consider the following setting: The performance requirements in the target market are assumed to be normally distributed, but the company has some uncertainty about the true mean θ and the true variance σ^2 of the distribution. The prior distributions, which will reflect the available prior knowledge, are chosen from the corresponding conjugate families in order to reduce the computational effort for determining the posterior market requirement distribution. Thus, as specified in Proposition 3.8, the prior mean follows a normal and the prior variance an inverse gamma distribution.

Let us further assume that the company estimates (e.g., based on experience from previous projects, expert opinion, etc.) the mean and the variance of the two conjugate prior distributions as specified in Table 3.1. Based on this estimation, the firm can now specify the prior distribution by determining the values of the corresponding parameters, i.e., μ, ν, α, and β.

Since $\pi(\sigma^2)$ is a $IG(\alpha, \beta)$ distribution, which has according to Appendix A.1.3 an expected value of $E(\sigma^2) = \frac{1}{\beta(\alpha-1)}$ and a variance of $Var(\sigma^2) = \frac{1}{\beta^2(\alpha-1)^2(\alpha-2)}$, the parameters of the prior distribution can then be determined as $\alpha = 2.8$ and $\beta = 0.28$.

With the help of the marginal distribution of θ (Corollary 3.3), we can now determine the values of μ and ν. From $\pi(\theta) = St(\theta|\mu, \frac{\nu}{\alpha\beta}, 2\alpha)$ with $E(\theta) = \mu$ and $Var(\theta) = \frac{\alpha\sigma^2}{\alpha-2}$ (see Appendix A.1.4) we obtain $\mu = 0$ and $\nu = 1$. With this, we have all relevant parameters to specify the prior market performance requirement distribution of $m(d) = St(d|0; 2.6; 5.6)$ as derived in Proposition 3.7. It has an expected value of $E(d) = 0$ and a variance of $Var(d) = 4$ (see Fig. 3.6).

Assume now that the company conducts an additional market study at a given point in time during the development process to update its initial beliefs. Suppose for this reason that the marketing department interviewed $n = 6$ potential key customers of the product. The result of this study reveals that on average the customers in this sample require a performance of $\bar{d} = 1.5$ with a spread of $\sum_{i=1}^{6}(d_i - \bar{d}) = 2$.

Based on these findings, we can now determine the parameters of the posterior distributions for the two unknown quantities θ and σ^2. From

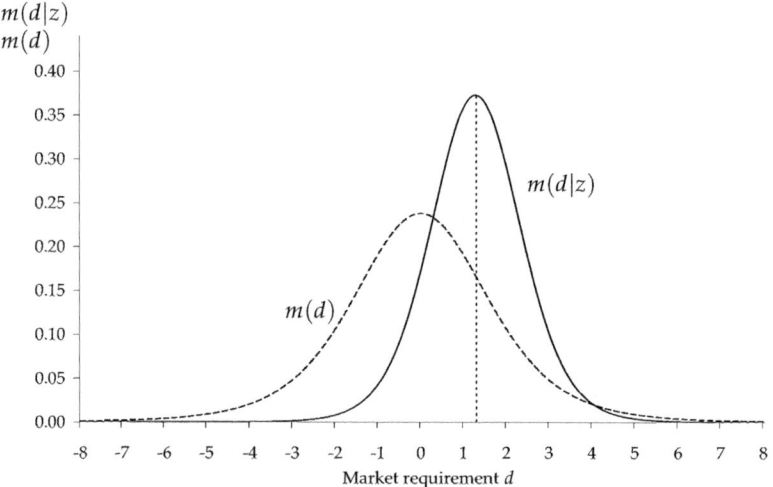

Fig. 3.6. Prior and posterior density distribution of market performance requirement

Proposition 3.8 it follows that $\alpha' = 5.8$ and $\beta' = 0.18$. With the help of the marginal posterior distribution of θ, $\pi_1(\theta|z)$, we can calculate the posterior mean and obtain $\mu' = 1.29$.

The posterior market requirement distribution is then t distributed, i.e., $m(d|z)$ is a $St(1.3; 0.01; 11.6)$ density, with an expected value of $E(d|z) = 1.29$ and a variance of $Var(d|z) = 1.33$. Fig. 3.6 shows the posterior density distribution. The density distributions for the mean and the variance of this example are provided in Appendix C.

3.2.4.3 Multidimensional Case

The multidimensional case of an unknown mean vector and an unknown covariance matrix is just the generalization of the univariate case specified in Proposition 3.8. As we can see in the subsequent proposition, it largely resembles the combination of the previously described multidimensional cases.

Proposition 3.9. *Suppose the performance requirements of the customers in the target market follow a normal distribution with unknown mean vector θ and unknown covariance matrix Ψ and $z = d_1, d_2, \ldots, d_n$ denotes a random sample of n independent (k-dimensional) observations of market performance requirements. Suppose further that the prior conditional distribution of θ when $\Psi = \psi$, i.e., $\pi(\theta|\psi)$, is a multivariate normal distribution with mean vector μ and covariance matrix vA ($\mu \in \Re^k$, $v > 0$, A is a $k \times k$ symmetric, positive-definite matrix) and the* distribution of the covariance matrix $\pi(\boldsymbol{\psi})$ is a inverse Wishart distribution with covariance matrix $\boldsymbol{\Sigma}$ and α degrees of freedom ($\boldsymbol{\Sigma}$ is a $k \times k$ symmetric, positive-definite matrix, $\alpha > k - 1$).

The posterior conditional distribution $\pi(\boldsymbol{\theta}|\boldsymbol{\psi}, z)$ is then a multivariate normal distribution with mean vector

$$\boldsymbol{\mu}^p = \frac{\boldsymbol{\mu} + nv\bar{z}}{1 + nv} \tag{3.92}$$

and covariance matrix

$$\boldsymbol{\psi}^p = \left[\left(\frac{1}{v} + n \right) \boldsymbol{\psi}^{-1} \right]^{-1}, \tag{3.93}$$

while the posterior distribution of $\pi(\boldsymbol{\psi}|z)$ is an inverse Wishart distribution with covariance matrix

$$\boldsymbol{\Sigma}^p = \left(\boldsymbol{\Sigma}^{-1} + \sum_{i=1}^{n} (d_i - \bar{d})(d_i - \bar{d})' + \frac{n}{1 + nv} (\boldsymbol{\mu} - \bar{d})(\boldsymbol{\mu} - \bar{d})' \right)^{-1} \tag{3.94}$$

and $\alpha + n$ degrees of freedom.

Proof. The proof follows in analogy to the proof of the univariate case as provided for Proposition 3.8. The likelihood function satisfies the following relation:

$$f_n(d_1, d_2, \ldots, d_n | \boldsymbol{\theta}, \boldsymbol{\psi}) \propto |\boldsymbol{\psi}|^{-1/2} \exp \left[-\frac{1}{2} \sum_{i=1}^{n} (d_i - \boldsymbol{\theta})' \boldsymbol{\psi}^{-1} (d_i - \boldsymbol{\theta}) \right]. \tag{3.95}$$

It follows from Eq. 3.37 and Eq. 3.62 that the following holds for the sum in the exponent of this relation:

$$\sum_{i=1}^{n} (d_i - \boldsymbol{\theta})' \boldsymbol{\psi}^{-1} (d_i - \boldsymbol{\theta}) = n(\boldsymbol{\theta} - \bar{d})' \boldsymbol{\psi}^{-1} (\boldsymbol{\theta} - \bar{d}) + tr(s\boldsymbol{\psi}^{-1}), \tag{3.96}$$

with

$$s = \sum_{i=1}^{n} (d_i - \bar{d})(d_i - \bar{d})'. \tag{3.97}$$

The joint prior distribution of $\boldsymbol{\theta}$ and $\boldsymbol{\psi}$ satisfies the following relation:

$$\pi(\boldsymbol{\theta}, \boldsymbol{\psi}) \propto |\boldsymbol{\psi}|^{-1/2} \exp\left[-\frac{1}{2v}(\boldsymbol{\theta}-\boldsymbol{\mu})'\boldsymbol{\psi}^{-1}(\boldsymbol{\theta}-\boldsymbol{\mu})\right]$$
$$\times |\boldsymbol{\psi}|^{-(\alpha-k-1)/2} \exp\left[-\frac{1}{2}tr(\boldsymbol{\Sigma}^{-1}\boldsymbol{\psi}^{-1})\right]. \quad (3.98)$$

It can further be shown that

$$\frac{1}{v}(\boldsymbol{\theta}-\boldsymbol{\mu})'\boldsymbol{\psi}^{-1}(\boldsymbol{\theta}-\boldsymbol{\mu}) + n(\boldsymbol{\theta}-\bar{\boldsymbol{d}})'\boldsymbol{\psi}^{-1}(\boldsymbol{\theta}-\bar{\boldsymbol{d}})$$
$$= (\frac{1}{v}+n)(\boldsymbol{\theta}-\boldsymbol{\mu}^p)'\boldsymbol{\psi}^{-1}(\boldsymbol{\theta}-\boldsymbol{\mu}^p) + \frac{n}{1+vn}(\boldsymbol{\mu}-\bar{\boldsymbol{d}})'\boldsymbol{\psi}^{-1}(\boldsymbol{\mu}-\bar{\boldsymbol{d}}) \quad (3.99)$$

and

$$\frac{n}{1+vn}(\boldsymbol{\mu}-\bar{\boldsymbol{d}})'\boldsymbol{\psi}^{-1}(\boldsymbol{\mu}-\bar{\boldsymbol{d}}) = tr\left[\frac{n}{1+vn}(\boldsymbol{\mu}-\bar{\boldsymbol{d}})'(\boldsymbol{\mu}-\bar{\boldsymbol{d}})\boldsymbol{\psi}^{-1}\right], \quad (3.100)$$

with $\boldsymbol{\mu}^p$ as defined above.

We know from Bayes' theorem that

$$\pi(\boldsymbol{\theta}, \boldsymbol{\psi}|d_1, d_2, \ldots, d_n) \propto f_n(d_1, d_2, \ldots, d_n|\boldsymbol{\theta}, \boldsymbol{\psi})\pi(\boldsymbol{\theta}, \boldsymbol{\psi}).$$

Under consideration of Eq. 3.99 and Eq. 3.100, the following holds for the joint posterior distribution:

$$\pi(\boldsymbol{\theta}, \boldsymbol{\psi}|d_1, d_2, \ldots, d_n) \propto \left\{|\boldsymbol{\psi}|^{-1/2} \exp\left[-\frac{1+vn}{2v}(\boldsymbol{\theta}-\boldsymbol{\mu}^p)'\boldsymbol{\psi}^{-1}(\boldsymbol{\theta}-\boldsymbol{\mu}^p)\right]\right\}$$
$$\times \left\{|\boldsymbol{\psi}|^{-(\alpha+n-k-1)/2} \exp\left[-\frac{1}{2}tr(\boldsymbol{\Sigma}^{p-1}\boldsymbol{\psi}^{-1})\right]\right\}, \quad (3.101)$$

where the expression inside the first set of braces resembles – when regarded as a function of $\boldsymbol{\theta}$ – the density function of a multivariate normal distribution with the mean vector $\boldsymbol{\mu}^p$ and covariance matrix $\boldsymbol{\psi}^p$ as specified above. The function inside the second set of braces is proportional to the density function of an inverse Wishart distribution with the covariance matrix $\boldsymbol{\Sigma}^p$ as defined above with $\alpha + n$ degrees of freedom.

3.2.5 Summary

In this section, we have presented the Bayesian updating formulations required for the derivation of our valuation model. Based on the assumption of normally distributed product performance requirements in the market, we have derived the corresponding prior and posterior (marginal) distributions (stemming from conjugate families) that address not only uncertainty about the mean, which is the in the literature prevailing case (cf. Sethi et al.

2005, p. 9 ff.), but also about the variance, and about both the mean and the variance. Besides the distributions for the one-dimensional performance parameter case, we have also derived the updating formulations for the multidimensional case which, to the best of our knowledge, has not been addressed by any decision model in the area of operations management yet. This extension ensures the applicability of our model to real-life investment problems and simultaneously addresses the criticism raised by Loch and Terwiesch (2005) regarding the limitation of the existing Bayesian updating models to product development settings due to the one-dimensional outcome space.

The assumed conjugate relationship between the prior and the likelihood distribution allows us to describe the updated market performance requirement probabilities through a known distributional class. Any other distribution could have been chosen for the prior and the likelihood as well. This would, however, require significant computational efforts for the determination of the posterior distributions. In addition, the posterior distributions for the different cases of uncertainty would not necessarily belong to a common class of distributions anymore. Thus, without appropriate conjugate priors, the Bayesian updating formulation would be quite complex which would solely complicate the analysis significantly without providing additional insights.

3.3 Information Updating Valuation Model

3.3.1 General Structure

Based on the considerations of the previous sections, we are now able to develop the actual model for our decision problem by integrating the derived Bayesian updating formulation into the presented basic decision model. In contrast to the basic framework, which models solely the technical development uncertainty, we will take a second source of uncertainty into account, namely the uncertainty about the true market performance requirements. Consequently, we need to be able to incorporate prior knowledge about the performance requirements that can be reduced by a mid-term information update in a Bayesian manner with data obtained, for example, from an additional market study. The optimal managerial response in the remaining development periods will then be adjusted to the updated market requirements in dependence on the actually realized technical performance level. By combining Bayesian updating with real options analysis, the model allows to determine the value of the project given the managerial flexibility to react to these two sources of uncertainty. We will further be able to quantify the value of the additional information on market performance requirements and to determine the optimal timing of such an update.

The problem setting is similar to the one of the basic model. The company faces the challenge to decide about an investment in a NPD project given the same managerial flexibility to react to technical development uncertainties. In contrast to the basic model, however, we assume that before the start of the project at $t = 0$, the company has besides technical uncertainty also some uncertainty about the true market requirement distribution. This results partly from the fact that customers often do not know themselves precisely at the start of a development project, which often has a duration of three and more years (cf. Cooper and Kleinschmidt 1994; Datar and Jordan 1997), what specific product performance they will need at the moment of the product launch. In addition, a perfect forecast is generally very expensive to obtain at this stage of the development process because it significantly requires more resources (e.g., manpower, earlier prototypes, expenses for additional market studies, etc.) than at a later stage in order to collect the necessary amount of information for an uncertainty reduction by the same level.

We therefore assume, as described in Section 3.2, that the company starts at time $t = 0$ with some prior information about the mean and/or variance of the market requirement distribution while being uncertain about its true shape. For the case that the mean θ ($\theta \in \Re$) as well as the variance σ^2 ($\sigma > 0$) is unknown, the market performance requirement at time $t = 0$ is then t distributed[41], i.e., $m(d) = St(d|\mu, \frac{\nu+1}{\alpha\beta}, 2\alpha)$. Thus, the expected terminal payoff is a priori given as

$$\Pi(x) = m + \Phi_{St}(x)(M - m), \tag{3.102}$$

where $\Phi_{St}(x)$ denotes the t distributed cumulative probability function of $m(d)$. The project value based on this initial market requirement information, denoted by $V_t^{prior}(x)$, can be determined with the given backward recursion (Eq. 3.7). Although this value addresses all a priori known uncertainties, it still ignores the possibility and hence, the value of a later update of these initial beliefs with information obtained from additional market studies. We will denote this additional information or signal by z, its cost by γ_t, and the updating point in time by τ.[42] Thus, the investment decision whether to start the project or not must be based on the project value which incorporates the value of a later market requirement update. As the later impact of the update is, however, a priori not known, the value of the project has to be determined in expectation of possible signal values.

---

[41]See Section 3.2.5 for details.

[42]Although the signal represents an entire data set, i.e., $z = d_1, \ldots, d_n$, we will – for ease of discussion – solely consider its most important moments, i.e., depending on the purpose of the update, the mean, the variance, or both.

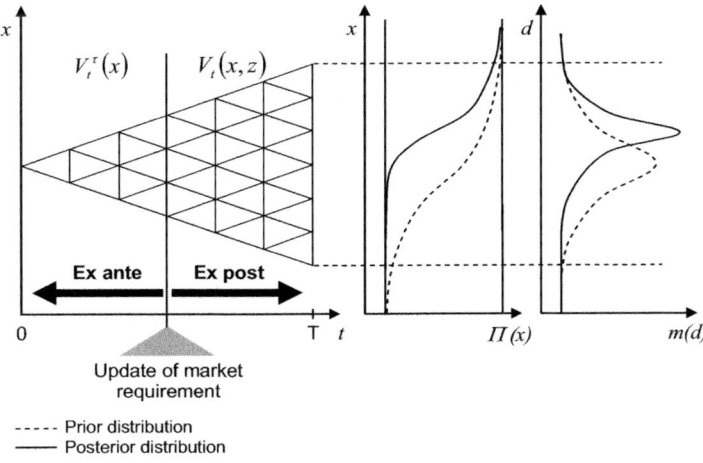

Fig. 3.7. Relevant perspectives for determination of project value

For the derivation of this expected project value, it is necessary to first define the project value given a certain signal z, i.e., $V_t(x, z)$, before deriving the project value in expectation of a later update at stage $t = \tau$, denoted by $V_t^\tau(x)$.[43] This distinction also can be explained considering two different perspectives (see Fig. 3.7), namely ex post and ex ante. The former determines the posterior project value $V_t(x, z)$, i.e., the project value after the signal has been obtained. The latter determines the expected project value $V_t^\tau(x)$, i.e., the project value in expectation of a later update. We will draw on this distinction in the subsequent sections.

3.3.1.1 Posterior Project Value

Once the signal about the latest performance requirements in the market has been obtained, the prior market requirement distribution can be updated. The resulting posterior distribution $m(d|z)$ determines the updated expected terminal payoff function of state x given the observed signal z:

$$\Pi(x, z) = m + \Phi_{St}(x, z)(M - m), \tag{3.103}$$

where $\Phi_{St}(x, z)$ denotes the posterior t distributed cumulative probability function of $m(d|z)$.

Based on this updated payoff function, management can now revalue the project and determine the optimal managerial response for the remaining

---

[43]To complete this listing, recall that $V_t(x)$ denotes the project value as determined with the basic model, i.e., without any updating possibility.

development stages. This is done in direct analogy to the project valuation of the basic model. Hence, the corresponding posterior project value that takes the obtained additional information into account can be formulated as a stochastic dynamic program (cf. Eq. 3.7) with the following value function:

$$V_t(x,z) = \max_{a_t} \left\{ -c_t(a_t) + \frac{1}{1+r} E_\omega \left[V_{t+1} \left(X_{t+1}(x,a_t,\omega_t), z \right) \right] \right\}. \quad (3.104)$$

It can be solved with the standard backward recursion.[44] Based on this definition, the posterior project value in expectation of an update can be derived next.

3.3.1.2 Expected Project Value

To determine the ex ante project value under consideration of managerial flexibility as well as the possibility of a market performance requirement update during the development process, we have to treat the signal as a random variable. We therefore assume that the signal z is a priori distributed according to a piecewise continuous density function f with a connected support $Z \subseteq \Re$. In other words, Z is the set of all possible values that the signal z may take. In absence of any additional information, the possible signal values are symmetrically centered around the mean of the prior distribution, i.e., $E[f(z)] = E(d)$. As explained before, this assumption is intuitive because in the presence of any additional information about a later market requirement drift, e.g., $E[f(z)] > E(d)$, one would incorporate this information already in the prior distribution and adjust the managerial actions correspondingly.

In order to solve the described problem of the ideal updating point in time, we have to determine the project value based on the expected impact of the information update. The value function of the corresponding stochastic dynamic program is given as:

$$V_t^\tau(x) = \begin{cases} -\gamma_t + E_z \left[V_t(x,z) \right] & t = \tau \\ \max_{a_t} \left\{ -c_t(a_t) + \frac{1}{1+r} E_\omega \left[V_{t+1}^\tau(X_{t+1}(x,a_t,\omega_t)) \right] \right\} & t < \tau, \end{cases} \quad (3.105)$$

where $\tau \in \{1,\ldots,T-1\}$ denotes the possible points in time of the information update and $E_z \left[V_t(x,z) \right]$ the expected project value with respect to the signal z. $V_t(x,z)$ is determined by the above presented standard backward recursion (Eq. 3.104). $V_t^\tau(x)$ thus represents the real option value of

---

[44]Note that $V_t(x,z)$ does not contain the expenses for obtaining the signal; they have still to be subtracted, i.e., $V_t(x,z) - \gamma_\tau(1+r)^{-(\tau-t)}$.

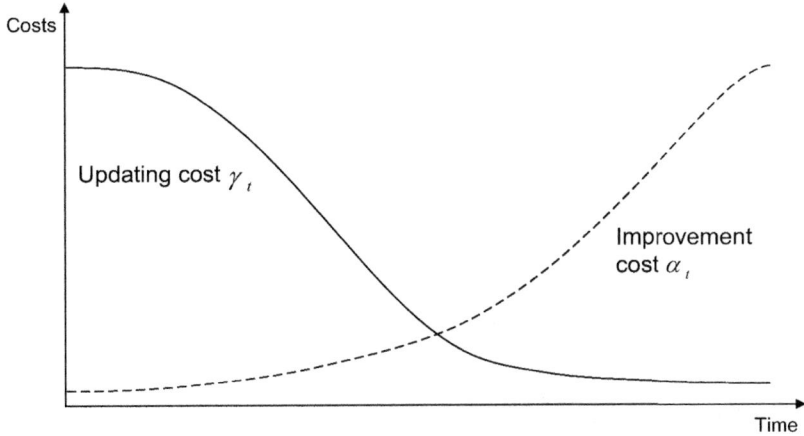

Fig. 3.8. Updating cost vs. improvement cost

the project that considers the optimal managerial response in expectation of a market requirement update at time τ.

We hereby assume that the information update can be conducted at any time $t = \tau$ during the development process. Solely the updating costs will decrease over time, i.e., $\gamma_{t_1} \geq \gamma_{t_2}$ $(t_1 < t_2)$, as depicted in Fig. 3.8. In other words, the cost for obtaining almost perfect information is the higher, the earlier the study is conducted. The intuition behind this assumption is that early in the development process much more effort is necessary, more customers have to be interviewed in more depth, more thorough studies are required, etc. in order to obtain the same level of information as immediately before the launch of the product when almost any customer knows what performance he actually requires. In addition, it is also the only possibility to model the fact that market studies generally become more precise towards the market launch of the developed product. The other conceivable assumption that the precision of the market studies increases over the development time is not feasible since real option valuation requires to consider all available information and hence, also this expected reduction of the market requirement variance, in the valuation of the project that is known at this point in time. The assumed cost structure therefore addresses this fact indirectly.[45]

The decision problem is therefore to decide whether to update the initial beliefs about the market requirement distribution early in the development process at a higher cost, but with the possibility to take the appropriate

---

[45]A similar assumption about the cost structure properties, for example, is made by Loch and Terwiesch (2005).

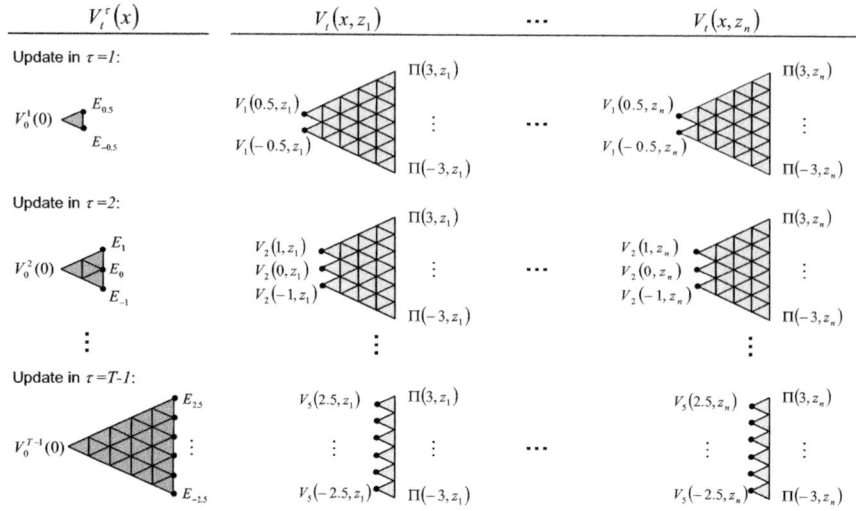

Note: For better readability of the illustration, we abbreviate $E_z\left[V_t(x,z)\right]$ by E_x and present only the update of a single parameter.

Fig. 3.9. Illustration of stochastic dynamic program for project value with update at time τ

counter measures early when these actions are still inexpensive and feasible or, to wait for a cheaper signal at the cost of a late and more expensive managerial response. Thus, management has to choose the optimal timing of the information update so that the corresponding present value of the project is highest. Details of this optimization problem are presented in Section 3.3.3.

Fig. 3.9 provides a schematic illustration of the dynamic program for a decision problem with a one-dimensional performance state space. For ease of representation, the possible expected signal values are assumed to be one-dimensional, representing either additional information about the market requirement mean or the variance, and their probability function is discretized. In such a case, the expected project value is determined as follows: $E_z\left[V_t(x,z)\right] = \sum_{i=1}^{n} V_t(x,z_i)p_i$ with $\sum_{i=1}^{n} p_i = 1$.

3.3.2 Value of Information

Without an update of the market performance requirements during the development process, the company will choose at each of the finally realized technical performance states of the project the managerial action which has initially been regarded optimal based on the prior information. However, since the market requirements will most likely have changed since the start of the project, the prior managerial policy will also have become outdated and will therefore be suboptimal. Hence, the expected project value based on the prior policy will be lower in absence of any updating costs than the one which considers the optimal managerial response to the latest information aggregated in the posterior market requirement distribution. To define a useful concept of the value of an information update, we have to determine the posterior as well as the expected project values based on the prior managerial policy first. For this derivation, it is insightful to maintain the previously introduced distinction between the two perspectives, namely ex post, i.e., when the value of the signal is known, and ex ante, i.e., when the project value can solely be determined based on the expected signal. We will start with the former perspective.

Let us denote the project value based on the prior managerial policy $a_{x,t}^{prior}$, i.e., the optimal managerial actions determined for the prior market performance requirement distribution, as $P_t(x)$. If we follow the notation introduced in the previous chapter, then $P_t(x, z)$ denotes the posterior project value (i.e., after the update of the market performance requirement distribution with signal z) when the (generally suboptimal) prior managerial policy $a_{x,t}^{prior}$ is applied. Its value can be determined with a backward recursion:

$$P_t(x,z) = -c_t(a_{t,x}^{prior}) + \frac{1}{1+r} E_\omega \left[P_{t+1}(X_{t+1}(x, a_{x,t}^{prior}, \omega_t), z) \right], \quad (3.106)$$

with the following terminal value as the starting point:

$$P_T(x,z) = \Pi(x,z). \quad (3.107)$$

On the other hand, we know from Section 3.3.1.1 that the project value, which considers the optimal managerial response to the updated market requirement distribution, is given as:

$$V_t(x,z) = \max_{a_t} \left\{ -c_t(a_t) + \frac{1}{1+r} E_\omega \left[V_{t+1}(X_{t+1}(x, a_t, \omega_t), z) \right] \right\}.$$

Thus, we can define the value of an information update similar to Marschak and Radner (1972, p. 85 f.) as the difference between the maximum project value given the optimal managerial response to the information update and the maximum project value one would achieve when applying the prior managerial policy. We will denote the value of the information update, or short value of information, by $V_t^I(x,z)$. For any time $t \geq \tau$, this is equivalent to

$$V_t^I(x,z) = V_t(x,z) - P_t(x,z). \quad (3.108)$$

Note, while the real option value is always greater than zero, the project value obtained based on the prior managerial policy does not necessarily have to be positive. If the optimal action in a certain performance state x_t is to abandon the project while it has been under the old market performance requirement forecast beneficial to continue or improve the project in this state, $P_t(x, z)$ is negative. In any case, however, $V_t(x, z) \geq P_t(x, z)$, so that the value of an information update is always positive, i.e., $V_t^I(x, z) \geq 0$.[46] The update is therefore from an ex post perspective valuable if the cost of the update is less than the value obtained from the new information through a better managerial response in the remaining development stages of the project, i.e., $\frac{\gamma_\tau}{(1+r)^\tau} \leq V_0^I(x, z)$.

However, if we want to determine a priori the value of the update, i.e., the expected value of information, we have to consider the project values based on the expected signal values. In analogy to the previous section, we will denote with $P_t^\tau(x)$ the expected project value for the at time τ updated market performance requirement distribution if the prior managerial policy $a_{x,t}^{prior}$ is applied. Its value can be determined as follows:

$$
P_t^\tau(x) = \begin{cases} E_z\left[P_t(x, z)\right] & t = \tau \\ -c_t(a_{x,t}^{prior}) + \frac{1}{1+r}E_\omega\left[P_{t+1}^\tau(X_{t+1}(x, a_{x,t}^{prior}, \omega_t))\right] & t < \tau. \end{cases}
\tag{3.109}
$$

Hence, the expected value of information $V_t^{I,\tau}(x)$, where τ indicates the updating point in time, is then the value of γ_t that makes Eq. 3.109 equal to Eq. 3.105. This is equivalent to the following equation:

$$
V_t^{I,\tau}(x) = \begin{cases} E_z\left[V_t(x, z)\right] - P_t^\tau(x) & t = \tau \\ \max_{a_t}\left\{-c_t(a_t) + \frac{1}{1+r}E_\omega\left[V_{t+1}^\tau(X_{t+1}(x, a_t, \omega_t))\right]\right\} - P_t^\tau(x) & t < \tau. \end{cases}
\tag{3.110}
$$

For any time $t \leq \tau$, the expected value of information is equivalent to the discounted expected value at the updating point in time τ, i.e., $V_t^{I,\tau}(x) = (1+r)^{-(\tau-t)}V_\tau^{I,\tau}(x)$.

Ex ante, the update of the market performance requirement is valuable if the expected value of the information update will exceed the corresponding cost for the update[47], i.e.:

---

[46]For a formal proof of this statement, see the proof of Proposition 4.11.

[47]The period in which the expected value of the information update will exceed the updating cost will be denoted by $\bar{\tau}$.

$$V_t^{I,\tau}(x) \geq \gamma_\tau (1+r)^{-(\tau-t)}. \tag{3.111}$$

But note, this rule is only the necessary condition for an update of the market performance requirements. It ensures that the information update has in expectation at least a value that compensates the expenses for the acquisition of the additional information; otherwise, the update – from an ex ante perspective – will be not valuable at the particular period and hence, would have to be rejected.

3.3.3 Optimal Updating Point in Time

Besides the primary issue of determining the potential value of the information update, the other interesting question for management is the one regarding the optimal timing of the update. As stated before, the managerial dilemma is either to conduct an early update when the signal costs are still high, but many development periods remain to take the appropriate counter measures or to wait until the information becomes cheaper at the expense of a limited managerial scope due to the reduced time-frame. In addition, the costs of counter measures generally increase over time.

Recall, $V_t^\tau(x)$ represents the posterior value of the project in expectation of a market requirements update at time τ. Hence, from the perspective of stage t, the optimal point in time for the update is the period for which the expected project value is highest (see Fig. 3.10)[48]. Mathematically formulated, the optimization problem is as follows:

$$V_t^{\tau^*}(x) = \max_\tau V_t^\tau(x), \quad 0 \leq t \leq \tau \leq T, \tag{3.112}$$

where τ^* denotes the optimal point in time for the information update. In case that there is more than one optimal updating point in time, one should update in the earliest possible period in order to have more flexibility to respond to contingencies.

The only case of practical interest, however, will be the one from perspective of $t = 0$, i.e.:

$$V_0^{\tau^*}(0) = \max_\tau V_0^\tau(0). \tag{3.113}$$

It is obvious that condition Eq. 3.111 must hold in any feasible state of this optimization problem.

---

[48]Note that this is only a schematic representation of the optimization problem. Depending on the other project parameters and cost functions, the expected posterior value function may have more than one optimum.

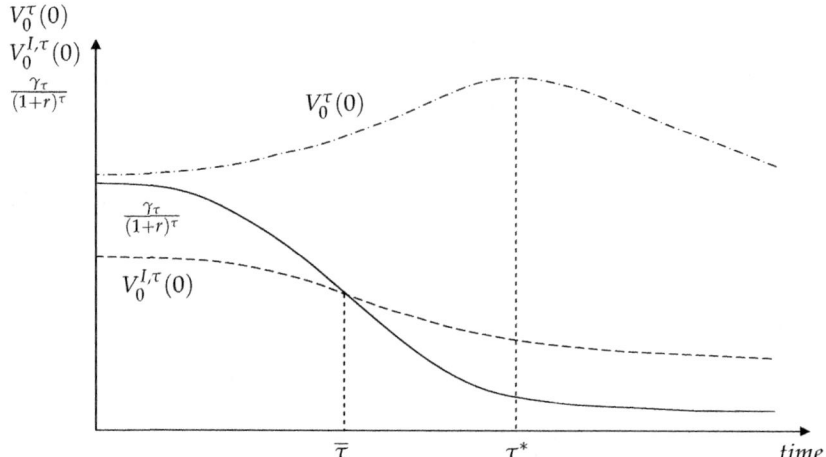

Fig. 3.10. Optimal updating point in time

3.3.4 Multidimensional State Space

The presented formulations of the information updating valuation model are based on a one-dimensional state space. In analogy to the previous sections, we will supplement the multidimensional case by sketching the most relevant considerations if k product performance attributes are considered. The multidimensional formulation of the model hereby follows in direct analogy the one of the basic model as derived in Section 3.1.5 and the multidimensional information updating formulations of Section 3.2.

Instead of a single performance parameter, the performance state space is here modeled by a vector X_t whose elements $X_1, \ldots, X_k$ represent the different technical specifications or performance parameters of the project. The corresponding market performance requirements are assumed to be multivariate distributed. For example, in case that management has uncertainty about the mean and the variance of the marginal distributions of the different performance parameters, i.e., uncertainty about the mean vector θ and the covariance matrix Ψ, the market performance requirements at time $t = 0$ are then multivariate t distributed, i.e., $m(d) = St(d|\mu, \frac{v+1}{\alpha\beta}, 2\alpha)$. Thus, a priori, the expected terminal payoff is given as

$$\Pi(x) = m + \Phi_{St}(x)(M - m), \tag{3.114}$$

where $\Phi_{St}(x)$ denotes the multivariate t distributed cumulative probability function of $m(d)$.

These initial beliefs can then be updated later during the development process with the observed signal $z = d_1, d_2, \ldots, d_n$ whose elements d_i contain market performance requirement observations on the k performance parameters considered in the model. The resulting posterior distribution $m(d|z)$ determines the updated expected terminal payoff function of state x given the observed signal z:

$$\Pi(x, z) = m + \Phi_{St}(x, z)(M - m),   \quad (3.115)$$

where $\Phi_{St}(x, z)$ denotes the posterior multivariate t distributed cumulative probability function of $m(d|z)$.

All derived value functions of the information updating valuation model for the one-dimensional state space, e.g., the posterior project value, the value of information, etc., can also be applied to the multidimensional case if the state variable x is replaced by the corresponding state vector x; all other variables have to be replaced by the corresponding vectors in the same way. For example, the function of the posterior project value which takes the observed signal z into account is now given as follows:

$$V_t(x, z) = \max_{a_t} \left\{ -c_t(a_t) + \frac{1}{1+r} E_\omega \left[V_{t+1} \left(X_{t+1}(x, a_t, \omega_t), z \right) \right] \right\}.   \quad (3.116)$$

Thus, the general setup of the model allows to conduct the same analyses as for the one-dimensional case if the derived multidimensional Bayesian updating formulation is applied. It simultaneously increases the practical applicability of the model significantly.

3.3.5 Summary

In this chapter, we have developed a general valuation model for NPD projects that takes the managerial flexibility into account to respond to two sources of uncertainty: The technical development risk as well as the uncertainty stemming from performance requirement changes of the customers. Both uncertainty sources influence the final payoff of the development project. By allowing for a reduction of the market requirement uncertainty through an information update at one point in time during the development process, we explicitly consider the value of additional information in the project value. With this approach, we not only extend the model of Huchzermeier and Loch (2001), but also develop the first decision framework that combines statistical decision theory in form of Bayesian analysis with a real options framework for NPD projects.

Our model further allows to determine the (expected) value of information which results from the adjusted managerial actions to the updated market requirement distribution. In case of an unforeseen change of the customer requirements, the managerial policy, which has been determined based on the initial estimates and beliefs, becomes outdated. If the counter measures of this policy are still applied in the realized performance states of the project at the different development stages, albeit the requirements of the market have changed, project value is destroyed either through unnecessary expenditures or missed opportunities. Finally, based on the defined range of possible signal values, the (in expectation) optimal updating point in time can be determined.

The key characteristics of the model and its managerial implications for decision making in NPD projects will be analyzed in the subsequent chapters.

4

Model Properties

Having developed a general valuation framework that combines Bayesian updating with real options analysis in the previous chapter, this chapter aims at analyzing the model and deriving properties in closed form. We will hereby study 1) how the project value changes if the uncertainty is reduced through an information update, 2) under which conditions it will be most beneficial to conduct an update, and 3), how the underlying cost structure influences the derived results. For the derivation of these properties it is insightful to maintain the distinction between the two perspectives (ex ante and ex post) as discussed in the previous chapter. We will start with an analysis of the corresponding properties of an information update once the signal is observed and hence, known. These results can then be applied to the analysis of the actual point of interest, namely model properties in expectation of a later information update.

Most of these properties can be derived in closed form. However, for some aspects, the model behavior is case dependent, requiring the consideration of project specific characteristics, e.g., the underlying cost structure. For these cases, we therefore resort to a numerical study that will be presented in Chapter 5. In addition, for ease of representation and in order to reduce the complexity of the proofs, we will limit our analysis to the one-dimensional performance parameter case. Due to the initially assumed a priori independence of the performance parameters, however, all properties and results remain valid for each particular dimension of the multidimensional case.

The chapter is structured as follows: We start with a brief discussion of the (basic) model characteristics and then derive the properties of a market performance requirement update based on an observed signal. Afterwards, we analyze the corresponding effects in expectation of the signal, i.e., the properties of the expected project value, the expected value of information, and the optimal timing of the update.

C. Artmann, *The Value of Information Updating in New Product,*
DOI: 10.1007/978-3-540-93833-0_4, © Springer-Verlag Berlin Heidelberg 2009

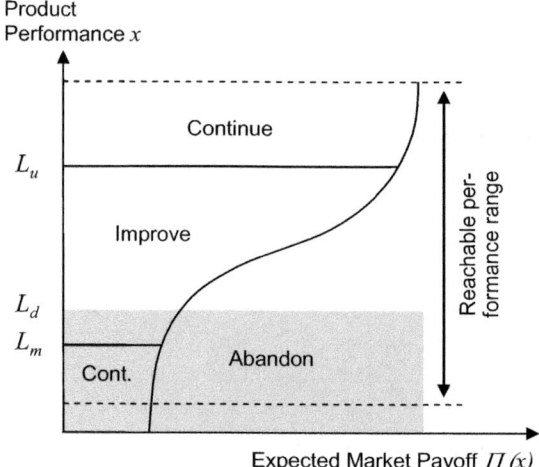

Source: Adapted from Huchzermeier and Loch (2001)

Fig. 4.1. Control limits of optimal managerial policy in basic model

4.1 Basic Properties

Our information updating valuation model, as presented in the previous chapter, builds upon an extended version of the real options valuation model developed by Huchzermeier and Loch (2001). For the following comprehensive analysis and discussion of our model, it is indispensable to elucidate some properties of the basic model. In this section, we will therefore briefly summarize the relevant properties of the former authors and derive other properties that have not been explicitly addressed yet, but which are important for the later interpretation of the results.

Huchzermeier and Loch (2001, p. 91) have characterized the optimal managerial policy for the presented stochastic dynamic program (cf. Eq. 3.7 of Section 3.1.4) by three control limits as depicted in Fig. 4.1.[49] Improvement is worthwhile in the middle where the convex-concave payoff function is at its steepest, i.e., $L_u(t) \geq x > L_m(t)$. If the technical performance either exceeds the upper or falls below the lower limit, i.e., $x > L_u(t)$ or $L_m(t) \geq x$, the potential payoff increase is too small in order to justify the improvement cost α_t and it is optimal to choose continuation. However, if the expected payoff is so low that it does not compensate the continuation cost c_t, it is optimal to abandon the project. This is always the case if $x \leq L_d(t)$. Note that

---

[49]Regarding the applicability of this proposition to our model, see the discussion in Section 4.2.1.2 and Appendix B.

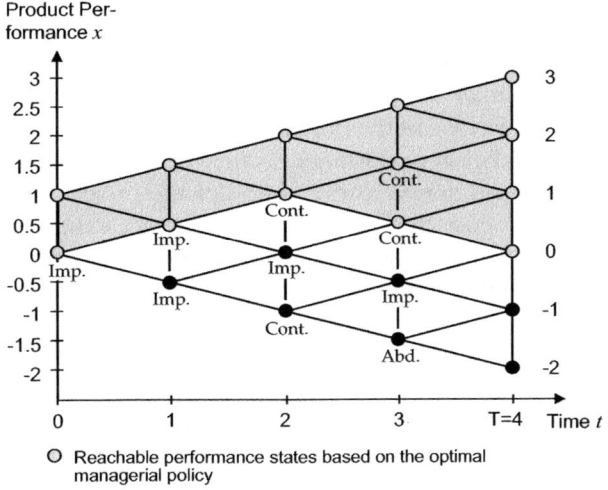

Product Performance x

O Reachable performance states based on the optimal
   managerial policy

Fig. 4.2. Example of reachable performance range under optimal managerial response

all control limits may be outside the range of the performance parameter x.[50]

This characterization of the model's optimal decision rule is important for the analysis of the underlying cost structure. Although we assume that the costs of the development project are given and hence, not a decision variable, it is still insightful for the later interpretation of the derived results to know how the underlying cost structure affects the project value. In particular, the value of managerial flexibility to respond to new information largely depends on the structure of the improvement costs α_t. To analyze the impact of these costs on the project value, we have to differentiate the performance states regarding their reachability. Whether a performance state x at stage t can actually be reached depends on the optimal managerial actions selected in the performance states of all previous stages. As illustrated in Fig. 4.2, only the highlighted performance states are reached if management selects the determined optimal managerial actions in every state. Thus, in the final stage T, performance states $x = -2$ and $x = -1$ are avoided if the optimal counter measures are taken in all previous periods $t = 0, \ldots, 3$.

If improvement is not the optimal managerial action for one of the performance states within the reachable range in a particular development stage, it is intuitive that an increase of the improvement cost of that stage

---

[50]For the presented model with the managerial option to improve the product performance by one level and a performance variability of $N = 1$, the range of the product performance x within the binomial lattice tree is $[-t/2, 3/2t]$.

does not affect the overall project value. In case of the illustrated project in Fig. 4.2, for example, an increase of the improvement cost at period $t = 3$ would not affect the project value $V_0(0)$ since improvement is only optimal in performance state $x_3 = -0.5$ which lies outside the reachable range. On the other hand, a decrease of the improvement cost may increase the project value since the number of performance states increases where improvement may be preferred over simple continuation of the project. Finally, the level of the continuation cost directly affects the project value, i.e., the higher the cost, the lower the project value. Simultaneously, the number of performance states where abandonment is the optimal managerial response might increase.

Although these just described insights are very intuitive, we will briefly summarize them in the subsequent proposition. The intention is to make the intuition of these properties mathematically more precise.

Proposition 4.1. *Consider a project which has a strictly positive project value, i.e.,* $V_0(0) > 0$. *In other words, if the optimal managerial actions are chosen in every performance state, there is a positive probability that the project will be launched in the market. Then, the following holds for changes of the underlying cost structure:*

1) The project value decreases if the continuation cost of any period is increased; 2) in case that improvement is the optimal managerial action for a state x at period t that lies within the reachable performance range, the project value increases if the improvement cost of stage t is reduced; 3) if improvement is not optimal for any state within the reachable range, an increase of the improvement cost does not affect the project value, while a reduction may increase the project value.

Proof. Consider two identical projects, $V_t^1(x)$ and $V_t^2(x)$, of which the second one has higher development costs at stage t, i.e., either $c_t^1 < c_t^2$ or $\alpha_t^1 < \alpha_t^2$.

In case of an increase of the continuation costs, the development costs of the first project in that stage are always lower than the one of the second project, i.e., $c_t^1(a_t) < c_t^2(a_t)$, – regardless of the optimal managerial action. Thus, by definition of the project value, $V_t^1(x) > V_t^2(x)$. Since the continuation costs accrue in any performance state, $V_0^1(0) > V_0^2(0)$.

Assume now that in both projects it is optimal to chose improvement in state $\bar{x}$ which lies within the reachable range. If $\alpha_t^1 < \alpha_t^2$ while the continuation costs are identical, i.e., $c_t^1 = c_t^2 = c_t$, we know that the following holds:

$$
\begin{aligned}
V_t^1(\bar{x}) &= -c_t - \alpha_t^1 + \frac{1}{1+r} E_\omega \left[V_{t+1} \left(X_{t+1}(x, a_t, \omega_t) \right) \right] \\
&> -c_t - \alpha_t^2 + \frac{1}{1+r} E_\omega \left[V_{t+1} \left(X_{t+1}(x, a_t, \omega_t) \right) \right] \\
&= V_t^2(\bar{x}).
\end{aligned}
\tag{4.1}
$$

Since $\bar{x}$ is within the reachable performance range, the corresponding value will be considered in the backward recursion and hence, $V_0^1(0) > V_0^2(0)$.

However, if improvement of state $\bar{x}$ is the optimal managerial action but that state is outside the reachable performance state, then it is obvious that an increase of α_t does not affect the overall project value, i.e., $V_0^1(0) = V_0^2(0)$ although $V_t^1(\bar{x}) > V_t^2(\bar{x})$. If, on the other hand, the improvement costs decrease, the following holds for the performance states where improvement is not the optimal managerial action:

$$
\begin{aligned}
&- \alpha_t^1 + \frac{1}{1+r} E_\omega \left[V_{t+1} \left(X_{t+1}(x,1,\omega_t) \right) \right] - \frac{1}{1+r} E_\omega \left[V_{t+1} \left(X_{t+1}(x,0,\omega_t) \right) \right] > \\
&- \alpha_t^2 + \frac{1}{1+r} E_\omega \left[V_{t+1} \left(X_{t+1}(x,1,\omega_t) \right) \right] - \frac{1}{1+r} E_\omega \left[V_{t+1} \left(X_{t+1}(x,0,\omega_t) \right) \right],
\end{aligned}
\tag{4.2}
$$

i.e., the difference between the value obtained under continuation compared to improvement increases if the improvement costs decrease. Thus, the improvement region increases, i.e., the corresponding control limit $L_u(t)$ increases while $L_m(t)$ decreases. Whether this leads to an increase in the project value depends on the actual extent of the cost reduction and the payoff structure of the project since the model considers only discrete stages. This completes the proof of the statements above.

For a given project setting, these results imply that the higher the increase of the improvement costs over time, the lower the value of managerial flexibility and hence, the lower the project value. An interesting question in this context is, whether it is beneficial to start with a more flexible, but also more expensive design, i.e., higher upfront investment cost c_0, that leads at the same time to lower cost for corrective actions at later stages of the development process, i.e., lower improvement cost α_t. The effects of such a change in the underlying cost structure on the project value can be summarized as follows:

Corollary 4.1. *Consider a project which has a strictly positive project value, i.e., $V_0(0) > 0$. Then the following holds for a reduction of the improvement cost at stage t ($t > 0$) by δ_t ($\delta_t \geq 0$), i.e., $\alpha_t - \delta_t$, at the expense of a corresponding increase in the investment cost at stage $t = 0$ by δ_0, i.e., $c_0 + \delta_0$:*

1) The project value $V_0(0)$ increases if the necessary upfront investment is less than the (in terms of time value of money) probability weighted improvement cost reduction at stage t, i.e.,

$$
\delta_0 < \delta_t (1+r)^{-t} \sum_{x=s}^{t/2} p(a_{x,t} = imp),
\tag{4.3}
$$

where s denotes the lowest reachable state at period t and $p(a_{x,t} = imp)$ the probability to reach the states where improvement is the optimal managerial choice.

2) The following holds for the upper bound of the upfront investment cost, δ_0 :

$$\delta_t(1+r)^{-t} = \overline{\delta_0} > \delta_0, \tag{4.4}$$

i.e., for a positive impact on the project value $V_0(0)$, the upfront investment cost can never exceed the corresponding time value of money of the improvement cost reduction.

Proof. We know from Eq. 3.7 that the project value at time $t = 0$, i.e., $V_0(0)$, can be determined by solving the corresponding value function in a backward recursive manner, i.e.:

$$V_t(x) = \max_{a_t} \left\{ -c_t(a_t) + \frac{1}{1+r} E_\omega \left[V_{t+1} \left(X_{t+1}(x, a_t, \omega_t) \right) \right] \right\}, \tag{4.5}$$

with $V_T(x) = \Pi(x)$.

The resulting optimal managerial actions determine a reachable performance range within the lattice tree. The project value $V_0(0)$ is only affected by an improvement cost reduction δ_t at stage t if the optimal managerial action (after the cost reduction) is "improvement" in at least one of the reachable performance states at this stage, i.e., $a_{x,t} = imp$. Hence, the probability to reach these improvement states determines the expected value of the improvement cost reduction at stage t, i.e., $\delta_t \sum_{x=s}^{t/2} p(a_{x,t} = imp)$. Discounted to $t = 0$, we obtain $\delta_t(1+r)^{-t} \sum_{x=s}^{t/2} p(a_{x,t} = imp)$ which corresponds to the increase of the expected project value $V_0(0)$. Thus, in order to be valuable, the upfront investment has always to be less than the corresponding improvement cost reduction at stage t.

For the better readability of the proof of the second statement, we denote the period of the cost reduction as t'. The probability of reaching the improvement states at period t' equals 100%, i.e., $p(a_{x,t} = imp) = 1$, only for the following case:

$$a_{x,t} = \begin{cases} \text{improvement} & \forall x \text{ in } t = t' \\ \text{continuation} & \forall x \text{ in } t = 1, \ldots, t' - 1. \end{cases}$$

If improvement or abandonment is the optimal managerial choice in any state of stage $t = 1, \ldots, t' - 1$, not all improvement states at stage t' are reached and hence, the expected value of the improvement cost reduction at time $t = 0$ is less than $\delta_{t'}(1+r)^{-t'}$. This implies that the upfront investment cost can be at most the corresponding time value of money of the improvement cost reduction and the second statement is proven.

The aspect of lower improvement costs at later development stages through a more flexible design approach upfront will be especially important in the presence of high initial market requirement uncertainty where management needs to be able to address the customer needs even at late stages. Generally, early changes as well as a broader design that allows for flexibility are less expensive than late modifications of the almost finished product.[51] The general rule of thumb states that the costs of design changes increase exponentially over time (e.g., Zangwill 1992). Nevertheless, a flexible design that allows for any contingencies of product performance adjustments at later stages does not necessarily have to be cheaper than late design changes. In this case, as the property above has shown, the higher upfront expenses for lower improvement costs at later stages are not compensated and thus, lead to a lower overall project value. But note, this proposition refers only to the basic model which solely considers the managerial response to technical uncertainty. Whether this paradigm changes in case of an uncertainty reduction possibility through a market requirement update will be analyzed in the next section. But before we turn to this issue, we will study at first the impact of an information update on the project value.

4.2 Properties for a Specific Signal

4.2.1 Posterior Project Value

In order to understand the characteristics of the developed valuation model, it is insightful to study the impact of a specific signal z on the project value, i.e., $V_t(x, z)$, in dependence on the achieved uncertainty reduction. The derivation of the corresponding properties will be in the focus of the subsequent analysis. The obtained insights will be helpful for the later analysis of the project value in expectation of an update, i.e., $V_t^{\tau}(x)$. In analogy to the derivation of the Bayesian updating formulation, we start with an isolated analysis of the base cases, a sole mean as well as a sole variance update, before discussing the impact of a simultaneous mean and variance update on the project value. Afterwards, we derive the properties for the value of information.

---

[51]This can be achieved, for example, through a modular product design or a development architecture that incorporates flexibility also in later product development stages, e.g., testing or manufacturing. For an overview of the latter development techniques, often also referred to as 'design for X' techniques, see Huang and Mak (1998), for example.

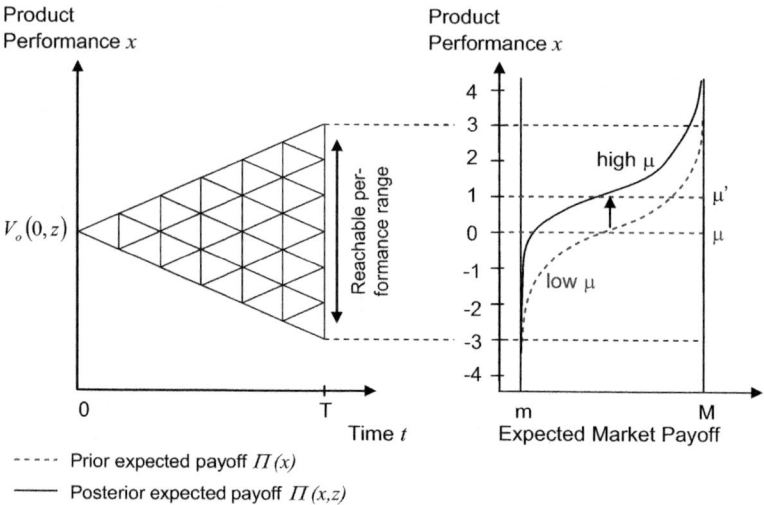

Fig. 4.3. Increased market performance requirement mean

4.2.1.1 Mean Change

The first interesting effect of a market performance requirement update on the project value is the one of a mean shift. Although we have seen in Section 3.2.2.1 that the update of the mean always simultaneously results in a variance reduction, the isolated analysis of a mean change is insightful for the interpretation of combined mean-variance effects. In the following, we will therefore study the impact of a pure, albeit hypothetical mean shift on the overall project value.

For the derivation of the subsequent properties, we will consider two projects that only differ in one parameter of their market performance requirement distribution while all other model parameters remain constant. All functions, parameters, and policies of the second project will be denoted with the upper bar. We will further – and slightly deviating from our present notation – denote with $V_t(x, \mu)$ the project value of performance state x at stage t given a market performance requirement distribution with mean μ. Since we analyze a sole mean update of the market requirement distribution, we denote the posterior project value by $V_t(x, \bar{\mu})$ for this case.

Intuition tells us that if the mean of the market requirement distribution becomes larger while all other parameters remain constant, more of the reachable performance states x of the project lie in the convex region of $\Pi(x)$ (see Fig. 4.3). In this case, the project value $V_t(x)$ decreases. More precisely, we can state the following proposition:

Proposition 4.2. *The project value $V_t(x)$ increases if μ, the mean of the market performance requirement distribution, decreases.*

Proof. We prove by backward induction. Suppose we have two identical projects except for a different market performance requirement mean μ and $\bar{\mu}$ with $\mu \leq \bar{\mu}$. By definition of the expected payoff function, $\Pi(x, \mu) \geq \Pi(x, \bar{\mu})$ and thus, $V_T(x, \mu) \geq V_T(x, \bar{\mu})$. Now assume that $V_{t+1}(x, \mu) \geq V_{t+1}(x, \bar{\mu})$. We need to show that $V_t(x, \mu) \geq V_t(x, \bar{\mu})$. Let the optimal action for the project with expected market requirement $\bar{\mu}$ be $\bar{a}^*$. Assume that the same action is applied for the project with expected market requirement μ. If $V_t'(x, \mu)$ denotes the corresponding project value under this assumption, we obtain:

$$V_t'(x, \mu) - V_t(x, \bar{\mu}) = \begin{cases} 0 & \text{if } \bar{a}^* = \text{abandon,} \\ \frac{1}{1+r} E\left[V_{t+1}\left(x + k(\bar{a}^*) + \omega_t, \mu\right) - & \text{if } \bar{a}^* = \text{continue} \\ \quad V_{t+1}\left(x + k(\bar{a}^*) + \omega_t, \bar{\mu}\right)\right] & \text{or improve.} \end{cases}$$

In addition, $V_{t+1}(x, \mu) \geq V_{t+1}(x, \bar{\mu})$ implies that

$$V_{t+1}\left(x + k(\bar{a}^*) + \omega_t, \mu\right) - V_{t+1}\left(x + k(\bar{a}^*) + \omega_t, \bar{\mu}\right) \geq 0. \qquad (4.6)$$

Note that the expected value of a non-negative random variable is non-negative. Thus, the following condition holds:

$$E\left[V_{t+1}\left(x + k(\bar{a}^*) + \omega_t, \mu\right) - V_{t+1}\left(x + k(\bar{a}^*) + \omega_t, \bar{\mu}\right)\right] \geq 0, \qquad (4.7)$$

and hence, $V_t'(x, \mu) - V_t(x, \bar{\mu}) \geq 0$. Since $V_t(x, \mu)$ denotes the project value under the optimal action a^* with a market performance requirement mean of μ while $V_t(x, \mu)$ considers only the suboptimal action $\bar{a}^*$, $V_t(x, \mu) \geq V_t'(x, \mu)$. Thus, it is obvious that $V_t(x, \mu) - V_t(x, \bar{\mu}) \geq V_t'(x, \mu) - V_t(x, \bar{\mu}) \geq 0$ and the statement above is proven.

The increase of the market requirement mean results not only in a lower project value, but also in a decrease of the project's NPV. As the payoff of the reachable performance range decreases with an upward mean shift, the static strategy of the NPV criterion to choose continuation in every performance state also results in a lower value. This simultaneous increase, however, makes it impossible to derive a general property of the option value $OV_t(x)$, i.e., it may in- or decrease if the market requirement mean increases. Thus, the portion of the managerial flexibility on the project value cannot be specified in general.

Applying the property about a mean shift to our information updating valuation model, we can derive the following property:

Corollary 4.2. *If the updating costs are negligible and the update of the market performance requirement distribution with signal z results solely in a mean shift while the variance remains (almost) constant ($\sigma'^2 \approx \sigma^2$), the following property of the posterior project value holds:*[52] *If the update increases the market performance requirement mean, i.e., $\mu' > \mu$, the posterior project value is lower than its prior value, i.e.,*

$$V_t(x, z) < V_t^{prior}(x). \tag{4.8}$$

Proof. Follows directly from Proposition 4.2.

In other words, an information update that results in an increase of the mean, i.e., the posterior mean is greater than the prior one, indicates that the initial estimation of the market performance requirement has been too optimistic and the determined prior project value has been too high. Thus, without an update, the company would have ignored the true performance requirement uncertainty in the market. This generally results in suboptimal managerial counter measures for the remaining reachable performance states of the project and hence, to a lower overall project value. In the worst case, the company could have even misleadingly adhered to the project, albeit it would have never been profitable at all.

4.2.1.2 Variance Change

The second case of a market requirement update that we study is the impact of a sole variance change on the project value. We know from Santiago and Vakili (2005) that an increase of the market performance requirement variability does not lead to an unidirectional result of the project value if the expected market payoff function is convex-concave.[53] Only for the case that the reachable range of the project falls completely either on the convex or concave part of the payoff function (Santiago and Vakili call these projects convex and concave projects, respectively), the following property can be derived:

Proposition 4.3. *For a convex project, if market requirement variability increases, then the project value increases. On the other hand, for a concave project, if market requirement variability increases, then the project values decreases.*

Proof. See Santiago and Vakili (2005) [Theorem 4.3].

---

[52]We will maintain the notation of Section 3.2 and denote with σ^2 the variance of the prior market performance distribution, while σ'^2 indicates the variance of the posterior distribution. The same notation holds for the mean.

[53]Santiago and Vakili assume a normally distributed market requirement. Since the t distribution of the market performance requirement in our model satisfies the same strict convex-concavity of the normal distribution, their theorem also holds for our case.

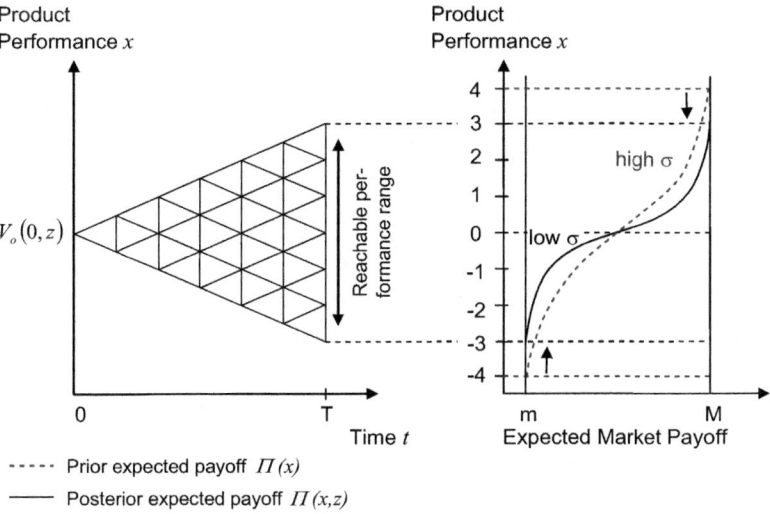

Fig. 4.4. Decreased market performance variability

However, if the variability of the product performance is limited to one performance state (i.e., $N = 1$)[54], as we have assumed for our model, we can show that an increase of the market requirement variance leads to a unidirectional result and a general property can be established (see Appendix B for details). In this case namely, Proposition 4 of Huchzermeier and Loch (2001) holds, which states the following:

Proposition 4.4. *The project value $V_t(x)$ decreases if σ, the market requirement variability, increases.*

Proof. See Huchzermeier and Loch (2001) [Proposition 4].

Fig. 4.4 illustrates the intuition behind this property: The greater the market performance variability, the greater the probability mass that lies beyond the reachable performance range of the development project, and hence, the lower the value of the project. This result is somehow surprising and counter intuitive compared with the standard real options theory which states that an increased variability results a higher option value as the managerial flexibility allows management to utilize the upside potential while limiting the downside risk (cf. e.g., Trigeorgis (1996, p. 4 ff.)). This general intuition holds, however, only for the case when uncertainty is resolved before a decision is made or any costs incurred. If uncertainty is reduced after all decisions are made, less variability leads – as the property above shows

---

[54]See Section 3.1.1 for details.

– to a higher value of managerial flexibility since more contingencies can be considered.

The NPV of the project, on the other hand, remains unchanged when solely the uncertainty is reduced since the mean and the expected payoff are not affected by the update. Thus, in the special case of a pure variance change, the increase of the option value is identical with the increase of the project value (or vice versa), which is the case when the market requirement variability decreases (increases). In the extreme case, when the variance is so large that there is no expected payoff difference between the reachable performance states, the value of managerial flexibility vanishes and the project value becomes equivalent to the NPV, i.e., the optimal managerial decision in every performance state is to continue the project.

Although we have a more general model formulation than the authors of the two cited papers above, their just described properties of the market requirement variability also hold for the properties of a pure variance update in our model. As derived in Section 3.3.1, the expected payoff in our model is given as $\Pi(x) = m + \Phi_{St}(x)(M - m)$. Since $\Phi_{St}(x)$ is strictly convex-concave increasing in x, the convex-concavity requirement of $\Pi(x)$, as assumed by Huchzermeier and Loch (2001), holds and their properties can be applied.

In particular, if we apply Proposition 4.4 to our valuation model, we can derive the following property about the impact of a sole variance update on the project value:

Corollary 4.3. *If the updating costs are negligible and the update of the market performance requirement distribution with signal z results solely in a variance change while the mean remains constant, i.e., $\mu' = \mu$, the following property of the posterior project value holds: If the update decreases the market performance requirement variance, i.e., $\sigma'^2 < \sigma^2$, the posterior project value is higher than its prior value, i.e.,*

$$V_t(x, z) > V_t^{prior}(x). \tag{4.9}$$

Proof. We know from Proposition 4.4 that $V_t(x)$ increases if σ decreases. Hence, a signal z which reduces the market performance variance increases the posterior project value, i.e., $V_t(x, z) > V_t^{prior}(x)$.

Whether the update of the market performance requirement distribution leads to a variance decrease depends of course on the observed additional data. Since the company in this case has uncertainty about the true variance, it will not be able to specify its prior value very precisely. Thus, the prior variance in this case will be rather high and the variance of the sample data will rarely exceed the initial beliefs. Otherwise the quality of the follow-up study could be doubted. For the more realistic former case, however, the just derived property holds in general, i.e., $Var(d) \leq Var(d|z)$ if $\sum_{i=1}^{n}(d_i - \bar{d})^2 \leq n Var(d)$.

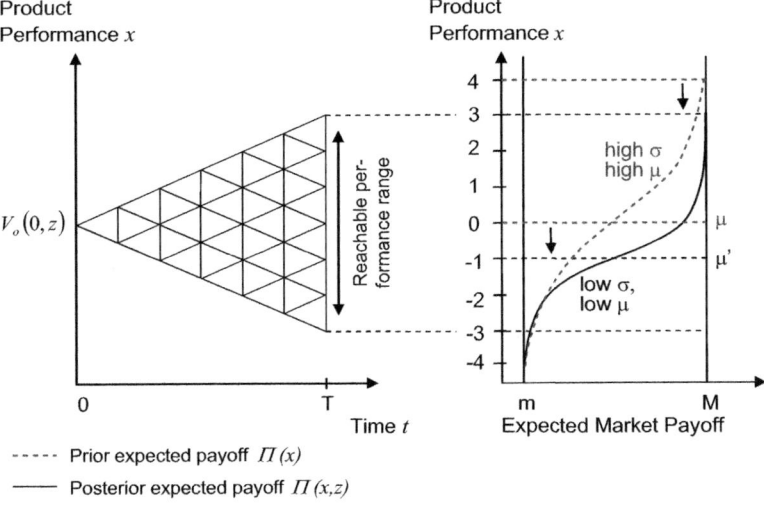

Fig. 4.5. Decreased market requirement mean and variance

4.2.1.3 Mean and Variance Change

The most general case is the one where the update of the market performance requirement distribution with signal z affects the mean as well as the variance. In contrast to the two previous scenarios, a simultaneous update of the mean and the variance does not necessarily lead to a unidirectional shift of the project value. Only if the mean as well as the variance decrease (see Fig. 4.5), we are able to state a general property:

Proposition 4.5. *The project value $V_t(x)$ increases if μ and σ^2, the mean and the variance of the market requirement distribution, decrease.*

Proof. We know from Proposition 4.2 that the project value $V_t(x)$ increases if the expected market requirement μ decreases. The same is true for an increasing market requirement variability (Proposition 4.4). Since the prior distributions of the mean and the variance are assumed to be independent, the effects are additive and the project value has the described characteristic for an opposed, but unidirectional simultaneous drift of the mean and the variance.

If, on the other hand, the mean increases while the variance decreases through the update of the market performance requirement distribution (or vice versa), only the following property can be derived:

Proposition 4.6. *If the increase in the mean is relatively small compared to the variance decrease, then there is a posterior variance $\overline{\sigma'}$ such that for all $\sigma' \leq \overline{\sigma'}$ the*

mean shift can be compensated and the project value remains constant or increases. A general unidirectional result for the project value when the mean increases while the variance decreases (or vice versa) cannot be established.

Proof. Consider two performance requirement density distributions with a single maximum (e.g., the normal distribution) which have the following moments: means μ and μ' ($\mu \leq \mu'$) and variances σ^2 and σ'^2 ($\sigma^2 > \sigma'^2$). Denote further with $\Pi(x)$ and $\Pi'(x)$ the corresponding expected payoffs and with $\bar{x}$ and $\underline{x}$ the highest and the lowest reachable performance state of the project, respectively. Then in case of the normally distributed performance requirement density, $\Pi'(\bar{x}) \geq \Pi(\bar{x})$ if $\sigma' \leq \overline{\sigma'} = \frac{\bar{x}-\mu'}{\bar{x}-\mu}\sigma$ (a similar condition can be derived in case of a t distributed density distribution). Simultaneously, the expected payoff of the lowest reachable performance state increases as well, i.e., $\Pi'(\underline{x}) \geq \Pi(\underline{x})$. Thus, if the mean and variance increase satisfy this condition, the expected payoff increases for all reachable performance states and we can conclude by definition (cf. Eq. 3.7) that the project value increases as well, i.e., $V_t'(x) \geq V_t(x)$. This proves the first statement of the proposition.

However, if $\sigma' \geq \overline{\sigma'} = \frac{\bar{x}-\mu'}{\bar{x}-\mu}\sigma$ (albeit $\sigma' < \sigma$), the increase of the mean cannot be compensated. In this case, the project value decreases. Hence, no general property can be derived.

The actual project value after a certain mean-variance change depends on the parameters of the model, in particular the underlying cost structure. Thus, for the analysis of such effects on the project value, we have to rely on a numerical study.

In addition, the above derived findings also apply to the option value of the project. It is obvious that under the just discussed updating scenario no conclusion can be drawn about the value of managerial flexibility. As discussed in Section 4.2.1.1, a sole update of the mean results in an increase of the project value as well as of the static NPV so that the option value may in- or decrease. Thus, in the more complex case of a simultaneous mean-variance change, no general insight about the value of managerial flexibility can be derived either.

If we apply the derived insights to our information updating valuation model, we can state the following property:

Corollary 4.4. *If the updating costs are negligible and the update of the market performance requirement distribution with signal z results in a unidirectional change of the mean as well as the variance, the following property of the posterior project value holds: If the update decreases the market performance requirement mean as well as the variance, i.e., $\mu' \leq \mu$ and $\sigma'^2 \leq \sigma^2$, the posterior project value is higher than its prior value, i.e.,*

$$V_t(x, z) \geq V_t^{prior}(x). \tag{4.10}$$

Proof. Follows directly from Proposition 4.6

As stated above, for all other updating scenarios, no general property can be established. In practice, however, the direction of the mean drift will be in most cases the decisive factor for the change of the project value. Moreover, the direction of the variance change is known since the uncertainty will be in every realistic updating scenario reduced, like in the before discussed case of a mean update, for example. In this particular case, the variance of the performance requirement distribution will always decrease through the update, regardless of the impact on the mean, i.e., $Var(d) = \sigma^2/n + \zeta^2 \leq \sigma^2/n + \sigma'^2 = Var(d|z)$, since $\sigma'^2 = \frac{\sigma^2 \zeta^2}{\sigma^2 + n\zeta^2}$.[55] The same holds for a simultaneous update of the mean and the variance if, as discussed in the previous section, the sample variance of the follow-up study does not exceed the prior variance.

In summary, the posterior project value $V_t(x, z)$ is the real option value of the project based on the updated expected payoff (determined from the with signal z updated market performance requirement distribution) under consideration of the optimal (adjusted) managerial response to the technical development uncertainty. The before discussed effects of the underlying cost structure on the project value remain valid regardless of the posterior market requirement distribution and hence, of the payoff function. Solely the magnitude of the cost structure change impact on the project value will vary in case of a market requirement update. But this can only be determined numerically.

The change of the project value due to the updated market requirement information, however, states nothing regarding the benefit of the information update. To determine the conditions under which an update of the market requirements is most valuable, we have to analyze the value of information. In addition, we will show that – counter to financial option pricing – investing more, e.g., in additional information to update prior beliefs, creates value. This runs counter to standard (real) option thinking.

4.2.2 Value of Information

The value of information is, as defined earlier, the difference between the project value obtained based on an optimal managerial response to the posterior market requirements and the one obtained if the prior managerial policy would have been applied. Thus, the larger the mismatch between the prior managerial policy and the optimal one, which is determined based on

---

[55]See Section 3.2.2 for details.

the updated market requirements, the larger the value of information. Regarding the impact of the uncertainty reduction on the value of information, we can state the following proposition:

Proposition 4.7. *Consider a project which has a strictly positive project value, i.e., $V_0(0) > 0$, where the information update (at the expense of updating cost γ_τ) of the prior market performance requirement distribution results in a posterior mean μ' and a posterior variance σ'^2 ($\sigma'^2 \leq \sigma^2$). Then $V_0^I(0, z)$, the value of information, increases if the posterior variance decreases; in other words, the higher the uncertainty reduction through the update, the higher the value of information.*

Proof. Consider two signals z_1 and z_2 that both result in a posterior mean μ' but different posterior variances σ_1^2 and σ_2^2 with $\sigma^2 \geq \sigma_1^2 \geq \sigma_2^2$. The corresponding values of information are denoted by $V_t^I(x, z_1)$ and $V_t^I(x, z_2)$.

First, assume that the market requirement mean remains unchanged through the update, i.e., $\mu = \mu'$. We know from Proposition 4.4 and the subsequent discussion that the benefit of the managerial options decreases as the market requirement variance increases, i.e., $V_t(x, z_1) \leq V_t(x, z_2)$. This simultaneously implies that the lower the posterior variance, the less suited will be the managerial actions based on the prior market requirement distribution $a_{x,t}^{prior}$. Hence, the mismatch between the project value based on the optimal managerial response and the one obtained based on the prior managerial actions increases as the posterior variance decreases, i.e., $P_t(x, z_1) \geq P_t(x, z_2)$.

Second, if the update – in addition to the variance reduction – also results in a mean shift, i.e., $\mu' < \mu$ or $\mu' > \mu$, this effect will be enforced since in this case the prior managerial policy will even be less suitable than before. Thus, the following holds:

$$\max_{a_t} \left\{ -c_t(a_t) + \frac{1}{1+r} E_\omega \left[V_{t+1}(X_{t+1}(x, a_t, \omega_t), z_2) \right] \right\} - P_t(x, z_2) \geq$$

$$\max_{a_t} \left\{ -c_t(a_t) + \frac{1}{1+r} E_\omega \left[V_{t+1}(X_{t+1}(x, a_t, \omega_t), z_1) \right] \right\} - P_t(x, z_1) \quad (4.11)$$

and, by the definition of the value of information, $V_t^I(x, z_1) \leq V_t^I(x, z_2)$. Hence, $V_0^I(0, z_1) \leq V_0^I(0, z_2)$, which makes the proof complete.

This result is particularly interesting for development projects with uncertainty about the market requirement mean. In the case of a pure mean update, the posterior variance depends, as mentioned before, solely on the ratio of the prior variance about the mean and the spread of the performance requirements among the customers in the target market. Hence, the uncertainty reduction can be easily determined upfront.

However, for a given uncertainty reduction, this property does not allow any inference about the dependence of the value of information on the prior uncertainty level, e.g., the lower the prior variance, the higher the value of information. Although, as seen in the previous section, the project value as well as the value of managerial flexibility will increase if the variance decreases, the mismatch between the prior managerial policy and the optimal one does not necessarily increase. The latter effect depends highly on the actual project setting. Even the opposite might be true for low prior variances. If the prior variance is already low and the signal primarily results in a variance reduction while the mean, for example, remains unchanged, the prior managerial policy might already be well suited for reaching the posterior market requirements. This does not need to be the case for a higher prior uncertainty even if the same uncertainty reduction is achieved. For this reason, we will study this effect in greater detail numerically.

Similarly, the value of the information update in dependence of the realized performance state cannot be determined in general in absence of the actual project data. In analogy to the characteristics of the optimal managerial policy with the described control limits (Section 4.1), an interesting aspect from a managerial point of view is whether the update is more valuable for certain performance states compared to others. Since an update of the market requirement distribution also changes the shape of the payoff function, the optimal managerial policy can only be determined in the presence of the actual model parameters. Moreover, the resulting project value functions, one with an optimal managerial response to the updated market requirements and one based on the prior managerial policy, might have – although being both convex-concave – a different shape. The value of information as the difference of these two convex-concave functions might therefore have any shape i.e., from a sole convex or concave to a multiple convex-concave shape with several inflection points. Thus, a general result how the value of information develops with respect to the realized performance state cannot be established.

The same holds true for the impact of the underlying cost structure on the value of the information update. Whether this value in- or decreases for an improvement cost reduction depends on the actual project setting. In most cases, the value of information will increase since lower improvement costs increase the scope for corrective actions after the update and by this, also the mismatch between the prior and the optimal managerial response to the (updated) market requirements. But this does not hold in general. Also in case of a cost structure change, no general result can be established. The value of information for two identical projects that only differ in their underlying cost structure might in- or decrease if improvement costs at a later development stage are reduced at the expense of higher upfront investment costs. In most instances, the value of information will increase for a certain cost reduction. Thus, updating gains in importance. However, whether such a cost structure change is valuable for the company solely depends on its impact on the project value. We will further elaborate on this issue in the next chapter.

Concluding, the derived model properties for an observed signal z illustrate the value of information as well as the impact of an information update on the project value after the signal has been observed. These insights are key for the analysis of the project valuation in expectation of a random signal and the determination of an optimal updating point in time which will be presented next.

4.3 Properties in Expectation of a Signal

The previous sections provided the basic information updating characteristics of our valuation model. In particular, the mean and variance effects as well as the optimal conditions for an information update could be derived from the analysis of the updated project value with an observed signal z. The remaining questions are 1) to which extent can these results be applied to the project value and the value of information in expectation of an update and 2) what properties can be derived for management to make optimal decisions for the project prior to its launch at time $t = 0$.

A priori, when the exact signal is not yet known, we have, in absence of any other information, to regard the signal as random. However, an updating possibility during the development process represents from an option's theory point of view already a value in itself. The conditions, when and under which project settings an information update will be most valuable in expectation of a later signal or the best operational strategies to respond to a later market uncertainty reduction during the development process will therefore be of particular interest from a managerial point of view. This will be the focus of the subsequent analysis.

4.3.1 Expected Project Value

The value of the project at any time before the update (i.e., $t < \tau$) of the market requirement distribution is determined as the average over all project values obtained from the possible posterior payoff functions. As stated before, the different posterior payoff functions are the results of an update of the prior market requirement distribution with all possible signals z out of the feasible and reasonable set Z. Since the density function of the signal is assumed to be symmetrically distributed around the prior mean, the corresponding project value is in this case an average of these opposing trends. The expected project value $V_t^\tau(x)$ will therefore generally differ from the finally realized one, $V_t(x, z)$. However, it represents the project value under consideration of the optimal managerial response to the two sources of uncertainty, namely the technical development and the market requirement uncertainty, through investments in additional resources and an requirement update based on the best available information at that time.

If the model properties of the previous sections are applied to the project value in expectation of an information update, we can state the following proposition:

Proposition 4.8. *Consider a project which has in expectation a strictly positive project value, i.e., $V_0^\tau(0) > 0$. Then $V_t^\tau(x)$, the expected project value, increases if the improvement cost α_t is reduced.*

Proof. Consider two identical projects which differ only in their underlying cost structure in the sense that the second one has lower improvement costs. To differentiate the two projects, we will denote all corresponding values of the second one with superscript c, e.g., $V_t^c(x)$.

First assume that only the improvement cost of the development stage after the updating point in time is reduced, i.e., $\alpha_{t_1}^c < \alpha_{t_1}$ $(t_1 > \tau)$. We know from Proposition 4.1 that a reduction of the improvement cost increases the project value, i.e., $V_t^c(x) \geq V_t(x)$. Since this property holds for every updated project value regardless of the observed signal, i.e., $V_t^c(x, z) \geq V_t(x, z)$ $(\forall z \in Z)$, also the expected project value with the lower improvement cost will always be at least as high as the one of the former project, i.e., $E_z[V_t^c(x, z)] \geq E_z[V_t(x, z)]$.

If the improvement cost is reduced at the stage of the update or in any stage before the update, Proposition 4.1 applies directly:

$$
\begin{aligned}
V_t^{\tau,c}(x) &= \max_{a_t} \left\{ -c_t(a_t) + \frac{1}{1+r} E_\omega \left[V_{t+1}^{\tau,c}(X_{t+1}(x, a_t, \omega_t)) \right] \right\} \\
&\geq \max_{a_t} \left\{ -c_t(a_t) + \frac{1}{1+r} E_\omega \left[V_{t+1}^{\tau}(X_{t+1}(x, a_t, \omega_t)) \right] \right\} \\
&= V_t^\tau(x),
\end{aligned}
\tag{4.12}
$$

and the statement above is proven.

In other words, low improvement costs are also in expectation of a later information update highly valuable. A reduction of the improvement costs at the expense of higher upfront investment costs is in case of a later updating possibility generally valuable under the same conditions as for the base case (without an updating possibility).

Corollary 4.5. *Consider a project which has a strictly positive expected project value, i.e., $V_0^\tau(0) > 0$. Then the following holds for a reduction of the improve-*

ment cost at stage t (t $\geq$ τ)[56] by δ_t ($\delta_t \geq 0$), i.e., $\alpha_t - \delta_t$, at the expense of a corresponding increase in the investment cost at stage t $= 0$ by δ_0, i.e., $c_0 + \delta_0$:

1) The expected project value $V_0^\tau(0)$ increases if the necessary upfront investment is less than the (in terms of time value of money) probability weighted expected improvement cost reduction at stage t, i.e.,

$$\delta_0 < \delta_t(1+r)^{-t} E_z \left(\sum_{x=s}^{t/2} p(a_{x,t}^z = imp) \right),$$ (4.13)

where s denotes the lowest reachable state at period t and $p(a_{x,t}^z = imp)$ the expected probability to reach the states where improvement is the optimal managerial choice at stage t (dependent on signal z).

2) The following holds for the upper bound of the upfront investment cost δ_0 :

$$\delta_t(1+r)^{-t} = \overline{\delta_0} > \delta_0,$$ (4.14)

i.e., for a positive impact on the expected project value $V_0^\tau(0)$, the upfront investment cost can never exceed the corresponding time value of money of the improvement cost reduction.

Proof. First, the expected project value $V_0^\tau(0)$ is only affected by an improvement cost reduction δ_t if at least one of the reachable performance states at stage τ in one of the resulting lattice trees (obtained through the possible signal values $z \in Z$) is affected by the improvement cost reduction δ_t at stage t, i.e., $a_{x,t}^z = imp$. The expected probability to reach these improvement states is therefore determined by the (in expectation) optimal managerial decisions prior to the update and the signal dependent optimal managerial decisions in the remaining periods τ to t. Applied to the improvement cost reduction δ_t and under consideration of the discount rate, we obtain $\delta_0 < \delta_t(1+r)^{-t} E_z \left(\sum_{x=s}^{t/2} p(a_{x,t}^z = imp) \right)$ which corresponds to the increase of the expected project value $V_0^\tau(0)$. Thus, in order to be valuable, the necessary upfront investment has always to be less than the corresponding improvement cost reduction at stage t.

For the better readability of the proof of the second statement, we denote the period of the cost reduction as t'. If the following holds for all states and all signals $z \in Z$ of an improvement cost reduction at stage t':

$$a_{x,t} = \begin{cases} \text{improvement} & \forall x \text{ in } t = t' \\ \text{continuation} & \forall x \text{ in } t = 1, \ldots, t' - 1, \end{cases}$$

[56]Note that for an improvement cost reduction at any stage before the update, i.e., $t < \tau$, the a priori unknown signal has not to be considered for the impact of the cost reduction on the expected project value. Thus, Corollary 4.1 applies directly.

then $E_z \left(\sum_{x=s}^{t/2} p(a_{x,t}^z = imp) \right) = 1$ since the possible signal values z are assumed to be symmetrically distributed around the mean of the market requirement distribution and hence, do not affect the expected value. As soon as one optimal managerial action changes, the expected value of the improvement cost reduction at time $t = 0$ is less than $\delta_{t'}(1+r)^{-t'}$. In other words, the upfront investment cost can be at most the corresponding time value of money of the improvement cost reduction in order to have no negative impact on the expected project value. This proves the second statement.

Lower improvement costs – even at the expense of a corresponding increase in the initial investment costs – always increase the managerial flexibility at later stages of the development project. Although the impact of such a cost structure change will be in most cases significantly higher for a project with a later updating possibility compared to one without it, i.e., $\Delta V_t^\tau(x) > \Delta V_t(x)$, the actual impact of such an effort depends on the particular underlying cost structure. Thus, the impact on the project value can in general only be specified as described above. We will therefore study the impact of a cost structure change on the expected project value in greater detail in the numerical study presented in the next chapter.

Besides the impact of the underlying cost structure on the expected project value, the dependency of the expected value of information on the updating point in time τ is of particular interest for management. In order to avoid trivial results, we assume throughout all remaining sections of this chapter that the cost of the information update is so small that there exists at least one point in time τ for which it is valuable to conduct the update, i.e., $V_t^{I,\tau}(x) \geq \frac{\gamma_\tau}{(1+r)^{\tau-t}}$ $(t \leq \tau)$.[57]

In expectation of the signal z, we can state the following proposition for the optimal updating point in time:

Proposition 4.9. *Assume that the updating costs are either negligible, i.e., $\gamma_t = 0$ $(t = 1, \ldots, T-1)$, or constant in terms of time value of money. Then the expected value of the project increases if τ, the updating point in time, decreases, i.e.,*

$$V_t^{\tau_1}(x) \geq V_t^{\tau_2}(x), \tag{4.15}$$

for $t \leq \tau_1$ and $1 \leq \tau_1 < \tau_2 \leq T-1$. In other words, the earlier the information update is conducted, the higher the expected project value.

Proof. Consider two identical projects with the same expected posterior values of the performance requirement distribution. We prove that the project with an update one period earlier than the second one has a higher expected value, i.e., $V_t^\tau(x) \geq V_t^{\tau+1}(x)$. We know from Eq. 3.105 that for $t = \tau$

---

[57] Such an assumption is common in the analysis of models. See, for example, Murto (2004).

$$V_t^\tau(x) = -\gamma_t + E_z[V_t(x,z)]$$

$$= -\gamma_t + \sum_{i=1}^{n} V_t(x,z_i)p_i$$

$$= -\gamma_t + \sum_{i=1}^{n} \max_{a_t}\left\{-c_t(a_t) + \frac{1}{1+r}E_\omega\left[V_{t+1}(x,z_i)\right]\right\}p_i$$

and

$$V_t^{\tau+1}(x) = \max_{a_t}\left\{-c_t(a_t) + \frac{1}{1+r}E_\omega\left[V_{t+1}^{\tau+1}(x)\right]\right\}$$

$$= \max_{a_t}\left\{-c_t(a_t) + \frac{1}{1+r}E_\omega\left[-\gamma_{t+1}\sum_{i=1}^{n}V_{t+1}(x,z_i)p_i\right]\right\}$$

$$= -\gamma_{t+1}\frac{1}{1+r} + \max_{a_t}\left\{-c_t(a_t) + \frac{1}{1+r}E_\omega\left[\sum_{i=1}^{n}V_{t+1}(x,z_i)p_i\right]\right\}.$$

However, since

$$V_t^\tau(x) = -\gamma_t + \sum_{i=1}^{n}\max_{a_t}\left\{-c_t(a_t) + \frac{1}{1+r}E_\omega\left[V_{t+1}(x,z_i)\right]\right\}p_i$$

$$= -\gamma_t + \sum_{i=1}^{n}\max_{a_t}\left\{-c_t(a_t) + \frac{1}{1+r}E_\omega\left[V_{t+1}(X_{t+1}(x,a_t,\omega_t),z_i)\right]\right\}p_i$$

$$\geq -\gamma_{t+1}\frac{1}{1+r} + \max_{a_t}\left\{-c_t(a_t)\right.$$

$$\left. + \frac{1}{1+r}E_\omega\left[\sum_{i=1}^{n}V_{t+1}(X_{t+1}(x,a_t,\omega_t),z_i)p_i\right]\right\}$$

$$= -\gamma_{t+1}\frac{1}{1+r} + \max_{a_t}\left\{-c_t(a_t) + \frac{1}{1+r}E_\omega\left[\sum_{i=1}^{n}V_{t+1}(x,z_i)p_i\right]\right\}$$

$$= V_t^{\tau+1}(x)$$

due to the positive pay-off structure of the managerial options, i.e., $E_\omega(\cdot) \geq 0$, and $\gamma_t = \gamma_{t+1}\frac{1}{1+r}$, we have proven the statement above.

This result is at first sight somewhat surprising because it seems that regardless of the underlying cost structure it is always optimal to update as early as possible. However, one has to keep in mind that this only holds true if the updating costs are constant in terms of time value of money. In this case namely, an earlier update incurs no higher expenditures, but provides management with more periods where optimal counter measures can be taken to respond to the latest market requirement information. In other words, the increase of the project value solely results from the remaining time to take the appropriate counter measures in response to the additional information. But note, in the presence of (in time value of money terms) non-constant updating costs, the ideal updating point in time may be at any stage $\tau = 1, \ldots, T - 1$. In addition, development projects that are identical except for their underlying cost structure may also have different optimal updating points in time.

Before we will analyze this in greater detail and derive a property for the ideal updating point in time under the consideration of updating costs, we first have to study the properties of the underlying necessary condition, namely the expected value of information. This will be done next.

4.3.2 Expected Value of Information

In analogy to the value of information, the expected value of information $V_t^{I,\tau}(x)$ corresponds to the value that management can expect at time t from a later update at τ ($t < \tau$) of the market requirements. As explained before, it is determined as the expected difference between the project value with an optimal managerial response to the possible posterior payoff functions and the one obtained based on the prior policy. The previously derived properties for the value of information also apply largely in expectation of a random signal.

If management has only uncertainty about the mean, the variance reduction through a possible update is predetermined by the ratio of the prior uncertainty about the mean and the market requirement spread in the population (sample variance). For this case, we can state the following property:

Proposition 4.10. *Suppose the market performance requirement variance is known while the mean is unknown. Then the following is true: the higher the uncertainty reduction through the update, the higher the expected value of information $V_0^{I,\tau}(0)$.*

Proof. Suppose we have two sets of signals, Z_1 and Z_2, whose elements result in the same posterior means, i.e., $\mu_1' = \mu_2'$, while the signals of the first set result in a higher posterior variance compared to the ones of the second set, i.e., $\sigma_1'^2$ and $\sigma_2'^2$ with $\sigma^2 \geq \sigma_1'^2 \geq \sigma_2'^2$. In other words, the signals of the second set result in a higher variance reduction.

We know from Proposition 4.7, however, that the value of information increases if the posterior variance decreases, i.e.,

$$\max_{a_t} \left\{ -c_t(a_t) + \frac{1}{1+r} E_\omega \left[V_{t+1}^\tau (X_{t+1}(x, a_t, \omega_t), z_2) \right] \right\} - P_t(x, z_2) \geq$$

$$\max_{a_t} \left\{ -c_t(a_t) + \frac{1}{1+r} E_\omega \left[V_{t+1}^\tau (X_{t+1}(x, a_t, \omega_t), z_1) \right] \right\} - P_t(x, z_1). \ (4.16)$$

Since the sets of potential signals differ only with respect to the uncertainty reduction, any pair of signal elements result in the same posterior mean and Eq. 4.16 holds. However, if each element of set Z_1 leads to a higher value compared to each corresponding element of set Z_2, the same is true for the expected values, i.e.,

$$E_z [V_t(x, z_2)] - E_z [P_t(x, z_2)] \geq E_z [V_t(x, z_1)] - E_z [P_t(x, z_1)], \quad (4.17)$$

with $z_1 \in Z_1$ and $z_2 \in Z_2$ and, by the definition of the expected value of information, $V_t^{I,\tau}(x, z_1) \leq V_t^{I,\tau}(x, z_2)$. Hence, $V_0^{I,\tau}(0, z_1) \leq V_0^{I,\tau}(0, z_2)$ which makes the proof complete.

The property also holds in case of uncertainty about the mean and the variance. However, it has no practical relevance anymore since the uncertainty reduction in this case is only known in expectation depending on the ratio of the expected posterior variance, determined over all possible scenarios out of the feasible set, and the prior variance.

The most interesting property, however, is the effect of the updating point in time on the expected value of information. We know from the considerations of the previous section that the earlier the updating point in time, the more stages and hence, the more review points are left where the company can select the optimal managerial options in order to respond best to the changed market requirements. It is therefore intuitive that – in absence of any updating costs – an early update is more valuable compared to a later one. Formally, we can derive the following proposition:

Proposition 4.11. *The expected value of information increases if τ, the updating point in time, decreases, i.e.,*

$$V_0^{I,\tau_1}(0) \geq V_0^{I,\tau_2}(0), \quad (4.18)$$

for $1 \leq \tau_1 < \tau_2 \leq T - 1$. In other words, the earlier the information update is conducted, the higher the expected value of information.

Proof. Consider two identical projects with the same expected posterior values of the performance requirement distribution. Assume further that these two projects have the same prior market requirement information and hence, the same prior managerial policies. Only their updating point in time differs. We prove that the value of information of the project with an update one period earlier is higher than the one of the second project, i.e., $V_0^{I,\tau}(0) \geq V_0^{I,\tau+1}(0)$.

We know by definition that the optimal managerial action chosen in state x at time t based on the posterior market requirement information is always at least as good as the actions of the prior policy. The posterior project value therefore exceeds the project value based on the prior policy, i.e.,

$$
\begin{aligned}
V_t(x,z) &= \max_{a_t} \left\{ -\gamma_t - c_t(a_t) + \frac{1}{1+r} E_\omega \left[V_{t+1}(X_{t+1}(x,a_t,\omega_t),z) \right] \right\} \\
&\geq -c_t(a_{t,x}^{prior}) + \frac{1}{1+r} E_\omega \left[P_{t+1}(X_{t+1}(x,a_{x,t}^{prior},\omega_t),z) \right] \\
&= P_t(x,z).
\end{aligned}
\tag{4.19}
$$

Thus, the value of information is always positive, i.e., $V_t^I(x,z) \geq 0$.

Based on these considerations, we can conclude the same for the expected value of information, i.e.,

$$
E_z \left[V_t(x,z) \right] \geq E_z \left[P_t(x,z) \right],
\tag{4.20}
$$

and

$$
\begin{aligned}
\max_{a_t} &\left\{ -c_t(a_t) + \frac{1}{1+r} E_\omega \left[V_{t+1}^\tau(X_{t+1}(x,a_t,\omega_t)) \right] \right\} \geq \\
&-c_t(a_{x,t}^{prior}) + \frac{1}{1+r} E_\omega \left[P_{t+1}^\tau(X_{t+1}(x,a_{x,t}^{prior},\omega_t)) \right],
\end{aligned}
\tag{4.21}
$$

so that $V_t^{I,\tau}(x) \geq P_t^\tau(x)$ and hence, $V_t^{I,\tau}(x) \geq 0$.

In addition, we know that the managerial actions chosen based on the prior information $a_{x,t}^{prior}$ are independent of the updating point in time τ. Thus, the value of $P_t^\tau(x)$ is independent of τ as well. In other words, $P_t^\tau(x)$ is so to speak constant with respect to the updating point in time τ.

However, we know from Proposition 4.9 that $V_t^\tau(x)$ increases if τ decreases so that, based on the definition of value of information (cf. Eq. 3.110), $V_t^{I,\tau}(x)$ and thus, $V_0^{I,\tau}(0)$ must be also increasing for a decreasing updating point in time and the statement above is proven.

The signal can be obtained at any time during the development process through an additional market study. At the moment of the information acquisition, the update of the market performance requirements provides the company with the opportunity to revalue its investment and adjust its optimal managerial actions. The crucial question for the management is therefore whether to update the market performance requirement early when there are many development periods left for corrective actions or to wait for a cheaper signal at the expense of a reduced time-frame for counter measures which are, however, at this point in time generally more expensive. It seems to be apparent that there is (at least) one point in time for which the update is more valuable compared to other points in time. Since the exact shape of the expected value of information function with respect to time depends on the actual project parameters, a general rule for the ideal updating point in time can, without the knowledge of the actual project data and updating costs, not be established.

We know, however, that lower improvement costs in a certain period increase the scope of corrective action, i.e., it is more probable that the higher payoff will compensate the additional expenses. A reduction of the improvement costs at the expense of higher upfront investment costs will therefore allow to postpone the update until the signal becomes cheaper. The lower improvement costs hereby provide the necessary flexibility to compensate for the shortened time-frame where counter measures can be taken. However, since this property depends on the actual project setting, we have to resort to the numerical study in order to derive the instances for which it will apply.[58]

4.4 Summary

With the conducted structured analysis in this chapter, we could derive a set of general model properties. We have shown that – for a specific and known signal – the posterior value of the project increases if the mean or the variance of the market requirement distribution decreases through the information update. This result also holds true for a simultaneous mean-variance decrease. For an opposed directed impact, i.e., the market requirement mean increases while the variance decreases (or vice versa), however, no unidirectional result can be established. Although an increase of the mean can be compensated by a corresponding reduction of the variance, the precise threshold value depends on the underlying cost structure of the project. Finally, the value of information for an observed signal increases with the uncertainty reduction that can be achieved through the update. These properties for an information update with a specific signal illustrate the basic characteristics of our valuation model and facilitate the derivation of the actual interesting properties about the project value as well as the value of information in expectation of a later update.

The analysis of the latter characteristics revealed that the value of information also increases in expectation of a later update with the uncertainty reduction that can be achieved with it as long as the updating costs are constant in terms of time value of money. In addition, the expected project

---

[58]See Section 5.2.2 for details.

value as well as the expected value of the information update decreases if the update of the market requirement distribution is postponed by one period because then the management has less development stages left for taking the appropriate counter measure to respond to the latest market information. As the discussion has shown, in the presence of a later updating possibility, the structure of the improvement costs plays a decisive role in the flexibility to react to requirement changes of the customers. Up to which threshold an improvement cost reduction at the expense of higher upfront investment costs will increase the expected project value depends, however, on the specific project setting. This will therefore be studied in greater detail numerically in the next chapter.

5

Numerical Study

The analysis of our information updating valuation model in the previous section allowed us to derive several properties in closed form, e.g., the impact of a pure mean or a pure variance update, the effect of the uncertainty reduction on the (expected) value of information, and the role of the updating point in time on the expected project value. Other aspects, like the impact of a simultaneous mean-variance update on the project value or threshold values for a change of the underlying cost structure, turned out to be case dependent and hence, eluded a closed form analysis. For those properties, we will therefore resort to a numerical study in this chapter in order to derive further insights about our information updating model. Our objective is hereby 1) to illustrate the derived properties by studying a real NPD project example, 2) to study the impact of a general (i.e., both mean and variance) update of the market requirement distribution on the project value, 3) to analyze the effect of the update on the optimal managerial policy, and 4) to study the impact of the underlying model parameters on the expected project value and expected value of information.

The subsequent numerical analysis presents the results of a real NPD investment project faced by a semiconductor test equipment manufacturer. Since this business environment is highly competitive and dynamic, the project is exposed to numerous sources of uncertainty stemming from market, customer, and technical variabilities over the development time. Hence, the project success depends to a large extent on the successful resolution of uncertainty during the development process. Moreover, as explained below, the specific project characteristics allow reasonably simplification in order to adapt the complex reality to the model setup. These aspects make the selected case well suited for an application of our valuation model.

The insights obtained from the conducted comparative static analysis will have at first an illustrative character. It is in the nature of the study design that the derived results cannot qualify for generality. The complexity

C. Artmann, *The Value of Information Updating in New Product*,
DOI: 10.1007/978-3-540-93833-0_5, © Springer-Verlag Berlin Heidelberg 2009

of the model and its dependency on the underlying parameters prevents to study a certain property exhaustively by determining the corresponding values for all instances of every possible scenario. The reader should keep this in mind while studying the subsequent sections. However, this approach allows to study the change of the optimal managerial policy in presence of a later updating possibility compared to the one obtained with the presented basic valuation model of Huchzermeier and Loch (2001). In addition to facilitating the analysis of the other above mentioned properties, the grounding in a real investment project demonstrates the practical applicability of the derived valuation framework. But before turning to the numerical analysis and the discussion of the results, we provide the general setup of the numerical study.

5.1 Analysis Framework

5.1.1 Base Case

The numerical study will be conducted in a static comparative manner, i.e., starting from a base case whose underlying data will be later varied over a broad range of possible values. To ensure realistic proportions of the model parameters, we selected a real (ongoing) new product development project faced by a global manufacturer of semiconductor test equipment. Although the development process has been slightly simplified in order to adapt the complex R&D environment to our model requirements, the real-life example simultaneously shows the practical applicability of the developed model. For the sole purpose of analyzing and illustrating the model properties, however, any hypothetical example would serve the same purpose and could have been chosen instead. In the following, we will briefly explain the key characteristics of the selected development project, present the relevant data, and define the assumptions that are necessary in order to determine the project value under consideration of an updating mechanism with our model.

The selected project focuses on the development of a device that allows to measure the power supply of electronic components, e.g., microprocessors, memory chips, etc., as they are used, for example, in mobile phones. This product will be offered by the company as an additional card in their digital testers sold to manufacturers of such components. The key performance drivers of this product are the measurement range (e.g., of the voltage and the current) as well as the measurement capacity, i.e., the number of chips that can be tested simultaneously. As long as the required measurement range of the potential applications is met, the latter performance driver is the critical one for the customers because it determines directly

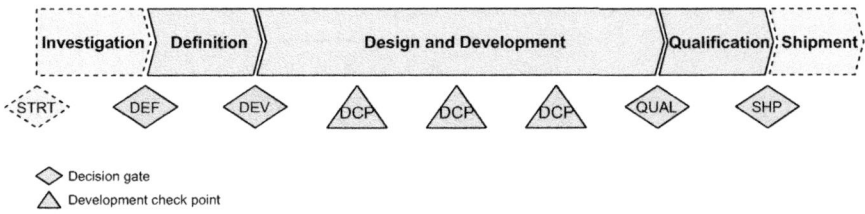

Fig. 5.1. Development process of power supply measurement device

their measurement and hence, their own production costs which again drive the profitability. In order to reduce the complexity of the numerical analysis, we limit the product performance to a single parameter and select the most important one, namely the measurement capacity of the device as the decisive performance parameter. This capacity is generally measured in the number of pins (in units of a thousand) that can be tested simultaneously.[59]

The company has established a classical stage-gate development process (Fig. 5.1). It starts with an investigation phase containing all necessary tasks to develop a business case for the product which has to demonstrate the profitability and the alignment of the product with the company's business strategy and strategic roadmap. If management approves the business case at the definition gate, the actual development process can be started with system-level designs during the definition phase and detailed designs as well as extensive testing during the design and development stage. The qualification stage finally serves for the refinement of the fabrication and assembly processes. Once the development project has passed the shipment gate, the finalized designs and prototypes are sent to a contract manufacturer for production. At all gates and regular development checkpoints, the R&D team discusses the current project status with the management, evaluates the developed technical solutions, and decides about current issues that need to be addressed in the further development of the product.

As the characteristic of the selected development project requires to start with design activities already in the definition phase due to the test intensive nature of the product, we consider the entire development process from the definition until the shipment gate for our model. Based on the development schedule of the project, we can almost perfectly approximate the development process with six equidistant development stages (i.e., $t = 0, \ldots, 5$), each with a duration of six months.

---

[59]Each chip has a certain number of pins with which it is connected to the circuit board. Thus, the number of chips that can be measured with this device results from the total capacity (in number of pins) and the number of pins per chip.

Table 5.1. Continuation and improvement costs of base case

Time t	0	1	2	3	4	5
Cont. cost c_t	2	3	15	18	24	28
Impr. cost α_t	4	6	22	25	29	42

The project length of six stages provides us, based on our model assumptions, with a range of seven reachable performance states in the last period.[60] At each stage, management has the choice among one of the three described managerial actions: abandon, continue, or improve. The latter option is limited to a shift by one performance state. The continuation of the project as well as the improvement of the product performance at a certain development stage incur the estimated expenditures as specified in Table 5.1. These data as well as all other costs and estimated revenues have been disguised for reasons of confidentiality; the proportions and dimensions, however, remain realistic. The significant increase in both types of costs in the third period results from the manufacturing and testing costs of the prototypes and the additional specialists required for these tasks. In addition, the project requires an upfront investment of $I = 10$ to put the necessary infrastructure in place, develop the business case, and conduct the market studies during the investigation phase.

While the development costs of such testing devices can generally be estimated with high precision based on past project experience, the key uncertainty for the valuation of the project are the expected revenues. The payoff largely depends on the finally realized technical performance of the product – in our case, on the measurement capacity – which has to meet the requirements of the customers. We approximated the market requirement for the measurement capacity with a normal distribution. Based on market studies and reviews of the product proposal with key customers, management estimated that the expected capacity requirement will be from today's perspective $E(d) = 150,000$ pins with a standard deviation of $\sigma_{prior} = 30,000$ pins.[61] The expected requirement of the customers also set the basis for the performance target of the project, i.e., $E(x) = E(d) = 150,000$ pins.

For the valuation of the base case, we assume that the estimated standard deviation represents the uncertainty about both the performance requirement mean ξ^2 and the spread of the market requirements among the

---

[60]Note that besides the "classical" $T + 1$ performance states, there is an additional one that can be reached due to the option to improve.

[61]Recall the following notational equivalence for the case of uncertainty about the true capacity requirement mean θ that will be studied later: $\theta = E(d)$ and $\sigma_{prior}^2 = Var(d) = Var(\theta) + E(\sigma^2) = \xi^2 + \sigma^2$. See Section 3.2.2.1 for details.

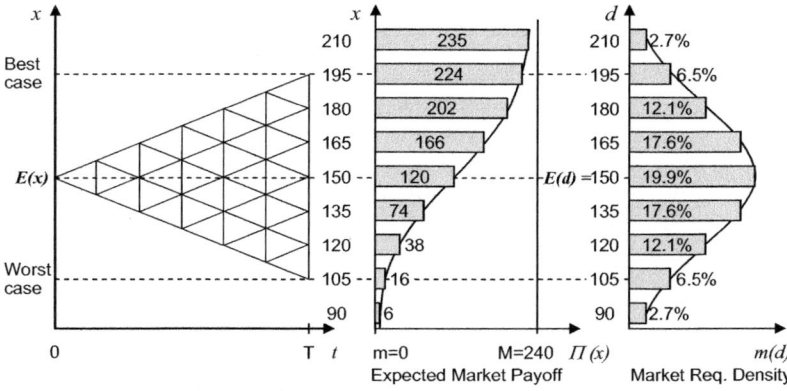

Fig. 5.2. Expected payoff and market requirement distribution

different customers σ^2, i.e., $\sigma^2_{prior} = \sigma^2 + \xi^2$. Depending on the products that are tested with this device and the market they are in, the customers have different capacity requirements. This variance is, however, generally either known or can be well estimated from past project experience or currently sold products. The greater source of uncertainty in this particular industry is the actual mean of the market requirements. Even detailed discussions with key customers about the project proposal before the start of the development project do not result in precise forecasts of the actual market requirements. The key problem in this dynamic business environment is that three years ahead even leading customers do not precisely know themselves which performance requirement they will actually need. An update of the expected measurement capacity during the development process is therefore essential.

We further assume that the technical performance x is also normally distributed around the mean $E(x)$ over the reachable range of performance states. Experts of the development team estimated that the highest product performance, which can be technically reached with the given development budget (i.e., without improvements), would be around $195,000$ pins if everything during the development process works out perfectly, while only a performance of $105,000$ pins may be realized in the worst case. The probabilities of these two adverse scenarios were regarded as significantly less than five percent; the standard deviation of the product performance was estimated around $15,000$ pins (Fig. 5.2). In order to avoid dealing with these large numbers all the time, we normalize the measurement capacity values around a mean of $\mu = 0$ which results in a corresponding prior market requirement standard deviation of $\sigma_{prior} = 2$ for this case (Fig. 5.3).

Table 5.2. Values for exogenous variables of base case

Description	Variable	Value
Development periods	T	6
Maximum payoff	M	240
Minimum payoff	m	0
Market requirement mean	μ	0
Market requirement std. dev.	σ_{prior}	2
Initial investment	I	10
Discount rate	r	6%

The price development of measurement devices, like the one considered here, is fairly stable so that the marketing specialists could derive a reliable estimate for the expected payoff range. The payoff ceiling was estimated to be $M = 240$ if the measurement device meets or exceeds the requirements of the customers; on the other hand, if the requirements cannot be met, the company will receive hardly any payoffs due to the strong competitive market environment. For ease of computation, we assume in this case a payoff of $m = 0$. Finally, all results are obtained using an arbitrary discount rate of $r = 0.06$ per period. Table 5.2 summarizes the parameter values of the base case.

In order to check the general profitability of the project based on the just provided data and to have a benchmark for the subsequent numerical analysis of our valuation model, we determine the project value as defined in the basic model (Section 3.1.4) at first. If we apply the presented backward recursion (cf. Eq. 3.7), which considers managerial flexibility, but not the possibility of a market requirement update, we obtain a project value of $V_0(0) = 48.3$. Fig. 5.3 illustrates the corresponding lattice tree of this investment example with the optimal value function and the optimal managerial policies for each performance state of the project. The project value $V_0(0)$ represents the compound real option value (i.e., the value of the managerial flexibility to choose improvement or abandonment in any stage) of the project in state $x = 0$ at stage $t = 0$ before the upfront investment cost I is deducted. Since the project value exceeds the initial investment cost of $I = 10$, i.e., the overall project value is with $V_0(0) - I = 38.3$ positive, it is worthwhile from today's perspective to invest in the project.

On the other hand, if we value the project based on the classical net present value criterion, we obtain $NPV = 1.3$. Thus, also on the basis of this static decision criterion, the project is profitable and would therefore be conducted regardless of the project inherent uncertainty. The fact that this value is significantly lower than the project value is not surprising since the NPV does not consider the managerial flexibility to respond to uncertainties

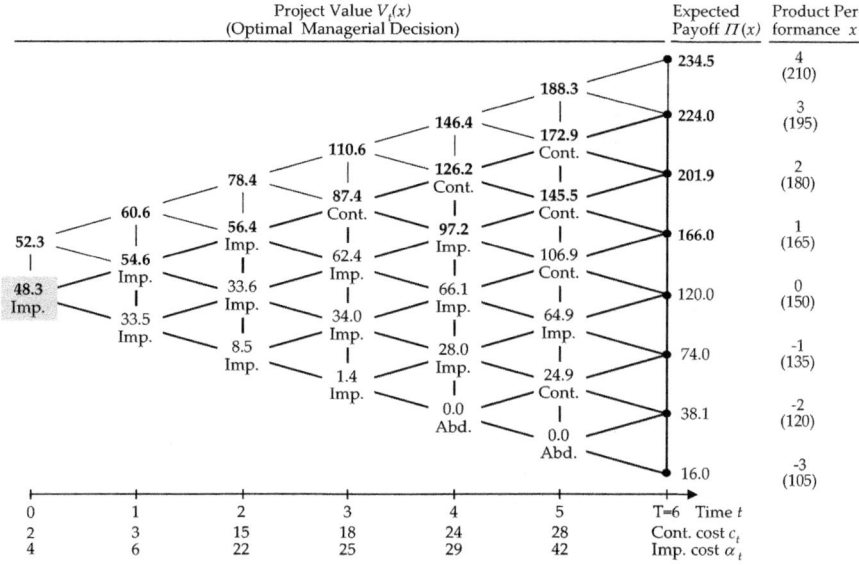

Note: Project values of reachable performance states appear in boldface type.

Fig. 5.3. Valuation of base case

during the development process; it corresponds to the special case where continuation is chosen in all states. With a decision on the basis of the NPV criterion, however, management underestimates the true value of the development project and may – misguided by this valuation result – wrongly reject additional investments or counter measures which would enhance the product performance and thus, the overall payoff of the project.

However, the determined (basic) project value $V_0(0)$ heavily depends on the estimated market data and neglects the possibility of a later update. Decisions made on this basis may therefore lead to wrong investments in counter measures during the development process if the actual market requirements change. Our developed valuation model explicitly considers this deficiency. But before determining the corresponding project value under consideration of an updating possibility and studying the model properties, we will briefly describe the setup of the numerical experiments.

5.1.2 Description of Experiments

The goal of the numerical study is, as explained before, to analyze and illustrate the properties of the project value for an update of the market performance requirement distribution during the development process as well as the characteristics of the value of the information update. The just described

Table 5.3. Parameter ranges of numerical study

Description	Variable	Value range
Spread of market requirements in target market	σ	$0.1, \ldots, 4$
Prior standard deviation of market requirement mean	ξ	$0.1, \ldots, 4$
Market requirement signal	z	$-4, \ldots, 4$
Updating point in time	τ	$1, 2, \ldots, 4$

NPD project hereby builds the basis for our experiments. For this purpose, we will analyze the characteristics of the model in the following by varying its key parameters over the ranges as specified in Table 5.3. The spread of the values of the different parameters have been set in such a way that they remain within realistic dimensions. In particular, we will study uncertainty scenarios that arise from different combinations of the prior variance (of the market requirement mean) and the sample variance. By allowing for standard deviations that range from $0.1, \ldots, 4$, we consider a much broader range than it will occur in any realistic project setting. Note also that a standard deviation of zero in any of the two uncertainties is not defined in the model and is therefore omitted. Starting with standard deviations of $\sigma = \xi = 0.1$ is sufficient for the analysis of the model properties since they are far below the variances of market data in real NPD projects. The same holds true for the upper limit.

Similarly, we will numerically study the model properties for signal values that range from $z = -4, \ldots, 4$. In case of an update of the prior market requirement mean, the signal values represent an observed or sampled average market performance requirement. With this range, we cover all relevant scenarios sufficiently. A signal value of $z = 4$ for example, may result – depending of course on its weight – in a posterior market requirement mean larger than 3, i.e., $\mu' > 3$, which already lies beyond the reachable range and indicates that the management has completely misjudged the market needs. Significantly larger signals and thus, posterior means are not representative for the assumed application setting of the model. If the market is so unpredictable that no realistic prior estimate is feasible, any upfront valuation of the project is worthless and thus, beyond the scope of the developed model. Finally, for the determination of the project value under consideration of a later updating possibility in expectation of a random signal, we assume, without loss of generality, that the signals are uniformly distributed around a mean of zero.[62] Fig. 5.4 illustrates schematically how the project value in

---

[62]This assumption has computational advantages. See Section 5.2.2.1 for further details.

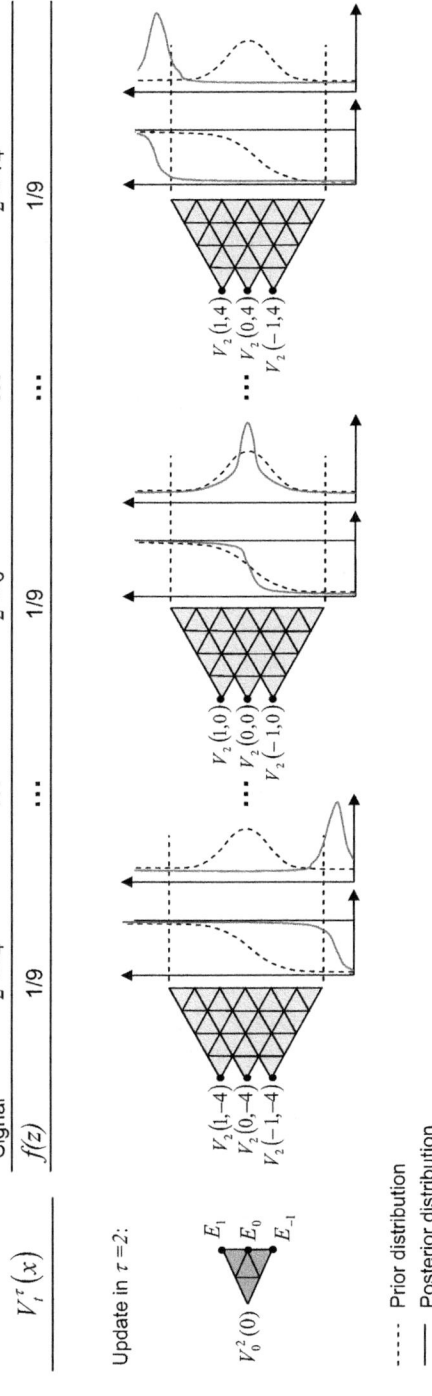

Fig. 5.4. Determination of project value in expectation of an update at $\tau = 2$

Note: For better readability of the illustration, we abbreviate $E_z\left[V_t(x,z)\right]$ by E_x and present only the update of the mean at time $\tau = 2$. $f(z)$ denotes the discretized uniform distribution of signal z.

expectation of an update at time $\tau = 2$ will be determined based on the defined parameter ranges.

We know from Section 4.2.1.3 that the update of the market performance requirement distribution in a Bayesian manner always results in a simultaneous variance change. This is even true if solely the mean is updated. For this reason, it seems to be sufficient to discuss the relevant effects on the project value on the basis of a pure mean update. Moreover, this approach also meets the market and industry characteristics of the selected development project, where companies have primarily uncertainty about the performance requirement mean while the spread among the customers can generally be well estimated. In order to reduce the complexity of our analysis, we do not study the impact of the (signal) sample size. This correlation is obvious from the Bayesian updating formulation provided in Section 3.2 and would only distract the attention from the actual points of interest.

To study the model properties for the selected parameter ranges, we will conduct a comparative static analysis (cf. Athey et al. 1998; Milgrom and Shannon 1994). By changing the underlying model parameters, e.g., the prior market requirement uncertainty or the sample variance, one at a time, it becomes apparent in what way the model results, i.e., the (expected) project value and the (expected) value of the information as well as the ideal updating point in time, depend on the underlying data. The impact of the exogenous uncertainty and thus, its reduction through the information update can be studied by determining the project values for an entire range of prior market requirement distribution parameters and signal values. The analysis of the impact of the underlying cost structure, e.g., the conditions and thresholds of an improvement cost reduction at the expense of higher upfront investment costs, however, is more complex. For the latter aspect, we have to study selected cost structure scenarios and compare their impact on the project and information updating value under different uncertainty settings. We will elaborate on this issue in Section 5.2.2.2.

To run the numerical experiments, we implemented the developed valuation model into a dynamic programming code (see Appendix D). This program was then solved for every instance of the considered scenarios in a backward recursive manner using the high-level technical computing language MATLAB, version 6.5 by *The MathWorks*, Natick, Massachusetts.

5.2 Numerical Results and Comparative Statics

In the subsequent presentation and discussion of the numerical results, we will maintain for reasons of consistency the distinction between the ex post and ex ante perspective of the project valuation as introduced in the previous chapter. This structure is helpful as the insights obtained about the

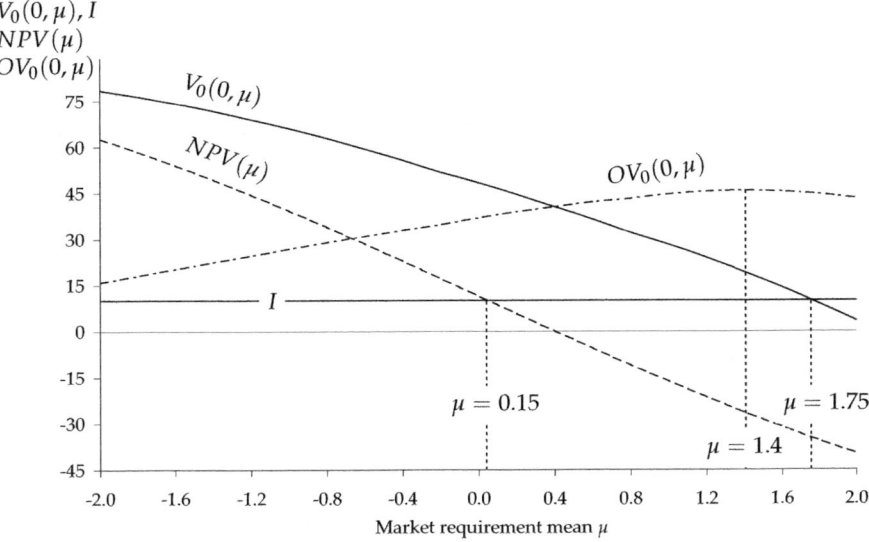

Fig. 5.5. Project value as a function of market requirement mean

project and information updating value for a specific (and known) signal facilitates the analysis and interpretation of the general case and actual point of interest, namely the model properties at time $t = 0$ in expectation of a random signal.

5.2.1 Analysis in Presence of a Specific Signal

5.2.1.1 Effects of a Mean or a Variance Update

We start our numerical analysis by illustrating the basic properties of an information update on the project value. Although we have derived the properties for a sole mean and a sole variance update in closed form, it is insightful to briefly illustrate the characteristics with the data of the base case. We focus at first on an update of the mean (at $t = 0$). As elucidated in Section 4.2.1.1, a pure mean shift is only of theoretical nature for an information update in a Bayesian manner since it always results in a simultaneous variance change. However, to illustrate the isolated effect of a mean change, we determine the project values for the base case data over a range of market requirement means. Fig. 5.5 depicts the results for the project value, the NPV, and the value of managerial flexibility (option value).

The project value $V_0(0, \mu)$ decreases as the market requirement mean increases (cf. Proposition 4.2). If the mean shifts towards the upper limit of the reachable performance range (i.e., $\mu > 1.75$), the project value falls below the initial investment cost $I = 10$ and the entire project loses its profitability. In this extreme case, when the market requirement mean exceeds the performance range of the development project, it is – even with improvements in all upper performance states – not possible anymore to ensure the profitability of the projects. Hence, a drastic shift of the market performance requirements cannot be compensated by improvements during the development process, but requires a re-definition of the entire project. This is also true for the case of a downward shift of the mean, where management has initially set the specifications of the product too high. Without an adjustment of the project target, there is a high probability that the development team strives for too expensive technical solutions and develops an over-engineered product which is too costly for the actual requirements of the customers and hence, reduces the potential payoff of the project.

These effects can also be seen by studying the NPV curve. We know from the valuation of the base case that the development project has a small positive NPV if the initial estimates hold, i.e., the market requirements of the customers are distributed around a mean of zero at the moment of the market launch. A slight increase of the requirements (i.e., $\Delta\mu = 0.15$) results in the fact that management would cancel the project, albeit it is still profitable under consideration of managerial flexibility. This is also evident from the shape of the option value curve: The considered managerial actions "improvement" and "abandonment" increase in importance as the market requirement mean increases up to $\mu = 1.4$. If, however, the market requirement mean reaches or even exceeds the upper limit of the technical feasible development range of the project, the available counter measures are not sufficient anymore to compensate for a payoff loss due to the requirement mean shift. The only feasible managerial action is to terminate the project in order to avoid further losses.

On the other hand, if the market requirement mean remains constant, i.e., $\mu = 0$, while solely the requirement variance increases, the project value decreases. The reason for this development is, as explained before, that part of the payoff mass escapes beyond the reachable range as the variance increases, i.e., the payoff function becomes flatter. Thus, the expected payoff difference over the reachable performance range decreases which also reduces the value of managerial flexibility responding to the technical development uncertainty. The NPV of the project is not affected since the mean remains constant. These effects are depicted in Fig. 5.6 which shows the corresponding value functions for a range of market requirement standard deviations. The value functions simultaneously show that the impact of a variance reduction on the project value is highest for intermediate uncertainties, where a variance reduction significantly increases the payoff mass within the reachable performance range. If the variance is, however, either extremely low or very high, an uncertainty reduction through an informa-

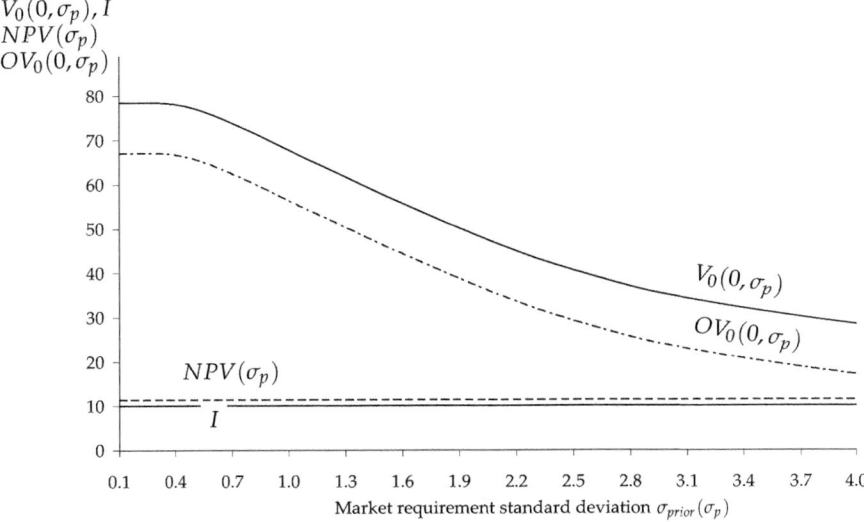

$V_0(0, \sigma_p), I$
$NPV(\sigma_p)$
$OV_0(0, \sigma_p)$

Fig. 5.6. Project value as a function of market requirement standard deviation

tion update will result only in a moderate project value increase since the payoff increase is diminishing (see also Fig. 5.7 in Section 5.2.1.2).

Note, in order to illustrate the impact of a pure variance change on the project value, as it would be the case of a sole variance update, we choose the same approach as for the illustration of the mean updating effect, namely to determine solely the project value for a range of reasonable standard deviation values. This approach is completely sufficient for the desired purpose.[63] After illustrating the model properties for a sole mean and a sole variance update, we will turn to the corresponding characteristics of a simultaneous mean-variance update in the next section.

5.2.1.2 Effects of a Simultaneous Mean-Variance Update

We know from Section 4.2.1.3 that a simultaneous change of the market requirement mean and the variance increases the project value in general only if the mean as well as the variance decrease. For an opposed directed change, the impact on the project value cannot be derived in closed form since it depends on the magnitude of the in- or decrease of the particular

---

[63]In order to completely model an update of the market requirement variance, we would have a priori to assume a specific (normal) distribution about the variance which then results in t distributed market performance requirements (see Section 3.2.3). However, this more formal approach would only distract from the desired purpose, namely to show the basic effects of a variance change.

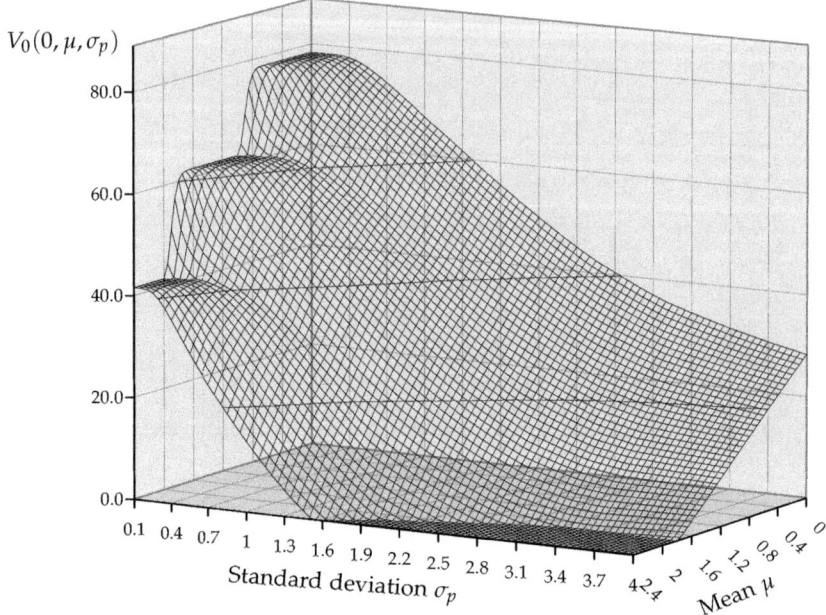

$V_0(0, \mu, \sigma_p)$

Fig. 5.7. Project value as a function of market requirement mean and standard deviation

parameters and on the underlying project data. To study the latter case, we maintain the approach of the previous section and determine the project value for a range of mean-variance combinations. Since the truly interesting case is the one when the mean increases while the variance decreases, we analyze solely the project value for mean values greater than zero. Fig. 5.7 plots the corresponding results over a mean range of $\mu = 0, \ldots, 2.4$ and $\sigma_{prior} = 0.1, \ldots, 4$ for the parameter data given in the base case.

It is apparent that the project value increases for a decrease in each of the two dimensions. The erratic increase of the project value for low variances results from the effect that the payoff function converges to a step function with mean μ as the market requirement variance diminishes. Since the model considers only discrete realizable performance states, the project payoff "jumps" in such a case for a small mean shift to the next level. However, this effect is primarily of theoretical nature since variances of this magnitude hardly occur in realistic project settings.

What can be seen on the basis of the level curves in Fig. 5.7 is that it requires a significant variance reduction in order to compensate for an increase of the mean. If, for example, the initial market requirement mean $\mu = 0$ increases to $\mu = 1$, the standard deviation of the market requirement distribution must simultaneously decrease from the initial estimate

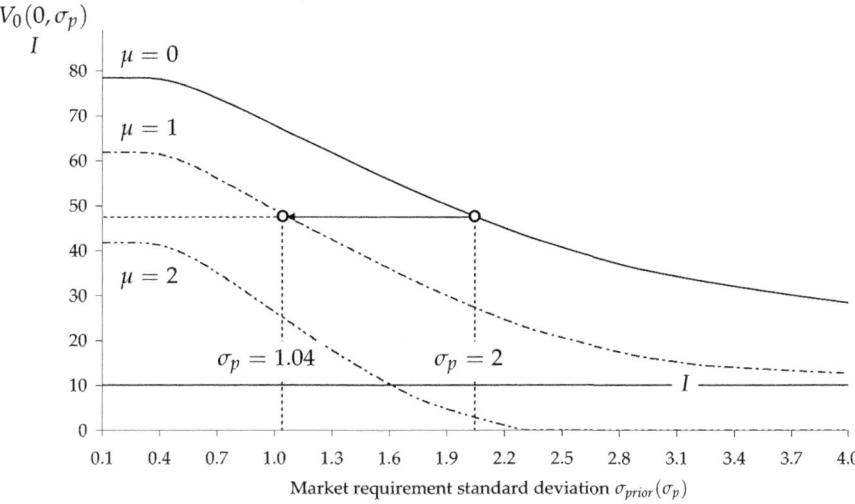

Fig. 5.8. Project value as a function of market requirement standard deviation at different mean levels

$\sigma_{prior} = 2$ to $\sigma_{prior} = 1.04$ in order to ensure the same project value (Fig. 5.8). A further increase of the mean to $\mu = 2$, however, can – under the given project setting – not be compensated anymore, i.e., the posterior project value falls below the initial one regardless of the magnitude of the variance reduction. Thus, only moderate mean shifts can accordingly be compensated by a simultaneous variance reduction.

So far, we have illustrated the potential effects of an information update by varying the variables of market requirement distribution which is sufficient for the desired purpose. In order to also demonstrate the Bayesian updating mechanism, we study the update of the prior market requirement distribution with a specific signal z in the following. Assume for this purpose that the performance requirements of the customers in the target market follow a normal distribution with an unknown mean θ and known variance σ^2, i.e., $d \sim N(\theta, \sigma^2)$. The prior estimate of θ is also assumed to be normally distributed, i.e., $\theta \sim N(\mu, \xi^2)$. Thus, the uncertainty at time $t = 0$ consists of the two described elements: the uncertainty about the true market requirement mean ξ and the requirement spread among the customers in the target market σ. We pick without loss of generality $\xi = 1.8$ and $\sigma = 0.9$ which results in a prior market requirement variance of $\sigma_{prior}^2 = 4$. The estimated expected market requirement (prior mean) remains unchanged, i.e., $\mu = 0$. For the sole purpose of illustrating this property, we further assume arbitrary updating cost at the beginning of stage $t = 1$ of $\gamma_1 = 5$.

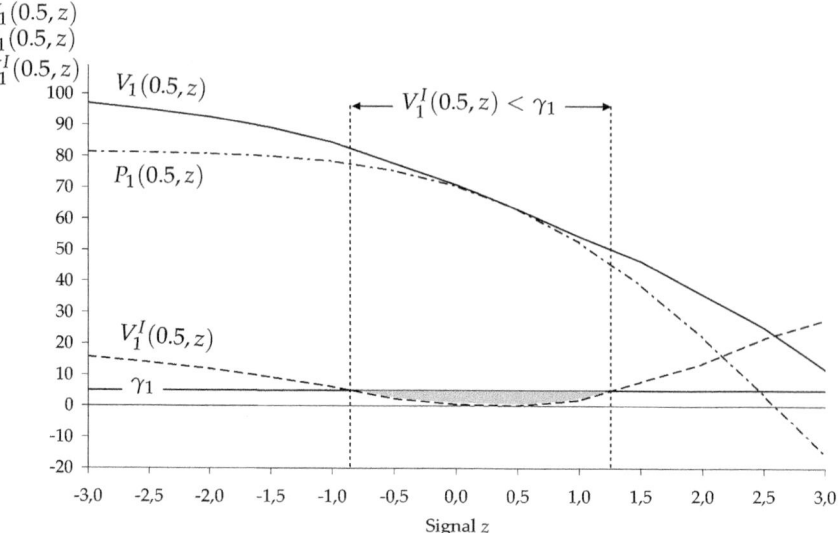

Fig. 5.9. Posterior project value as a function of signal z

If we determine the posterior project value for signal values ranging from $z = -3, \ldots, 3$, we obtain a value function as depicted in Fig. 5.9. Since we update solely the mean, we know from Section 3.2.2.1 that the posterior mean μ' is the weighted average of the prior mean μ and the observed signal z. With the above assumed uncertainty structure, the selected signal range corresponds to a posterior mean range of $\mu' = -2.4, \ldots, 2.4$. Simultaneously, the initial market requirement uncertainty $\sigma^2_{prior} = 4.05$ is reduced to a posterior standard deviation of $\sigma'^2 = 1.46$, which is equivalent to an uncertainty reduction by 64%.[64]

Assume further that the company reaches performance state $x = 0.5$ at stage $t = 1$. Then, depending on the observed signal z, the company will obtain $V_1(0.5, z)$ from the new product development project given an optimal adjusted managerial response at each review point. If management, however, adheres to its prior derived managerial policy while the actual market requirement shifts as indicated by the signal, it will obtain only $P_1(0.5, z)$. As long as the deviation is relatively small, the difference is negligible since it would not compensate for the updating expense γ_1. But especially for larger market requirement changes, information updating pays off as $V_1^I(0.5, z)$ clearly indicates. This shows that in this particular case an information update will also be from an a priori point of view beneficial when the actual

---

[64]It is measured as the reduction of the variance, i.e., $(\sigma^2_{prior} - \sigma'^2)/\sigma^2_{prior}$.

Table 5.4. Impact of update for projects with same prior variance ($z = -1$)

σ	ξ	$Var(d)$	$Var(d\vert z)$	Var Red.	$V_1(0.5,z)$	$P_1(0.5,z)$	$V_1^I(0.5,z)$
0.1	1.998	4	0.02	99.5%	95.76	81.38	14.39
0.2	1.990	4	0.08	98.0%	95.76	81.38	14.38
0.3	1.977	4	0.18	95.6%	95.48	81.38	14.10
0.4	1.960	4	0.31	92.2%	94.54	81.37	13.17
0.5	1.937	4	0.48	87.9%	93.10	81.29	11.82
0.6	1.908	4	0.69	82.8%	91.28	81.00	10.28
0.7	1.874	4	0.92	77.0%	89.13	80.42	8.70
0.8	1.833	4	1.18	70.6%	86.83	79.51	7.32
0.9	1.786	4	1.46	63.6%	84.24	78.24	6.00
1	1.732	4	1.75	56.2%	81.45	76.65	4.80
1.1	1.670	4	2.05	48.6%	78.51	74.78	3.73
1.2	1.600	4	2.36	41.0%	75.49	72.69	2.80
1.3	1.520	4	2.67	33.3%	72.50	70.45	2.05
1.4	1.428	4	2.96	26.0%	69.60	68.11	1.49
1.5	1.323	4	3.23	19.1%	66.73	65.72	1.01

realization of the signal z is unknown and the project can only be valued in expectation of the update.

In general, the value of the information update depends – in addition to the observed signal – on the underlying uncertainty structure of the project. We know from Proposition 4.7 that for a given prior market requirement distribution the value of information increases with the uncertainty that is reduced through the update. To illustrate this property, we determine the value of information for an update at stage $t = 1$ with signal $z = -1$ over a range of possible uncertainty structures of the project (Table 5.4).[65] All other parameters remain unchanged.

The results show that – from an ex post view – an update of the market requirement distribution is only beneficial if the prior variance can significantly be reduced. A relatively small reduction yields only a small additional payoff which will hardly cover the expenses for conducting the information update. In particular, if the prior variance of the market requirement mean is less than or equal the spread of the performance requirements among the customers, i.e., $\xi \leq \sigma$, the benefit of additional market studies and thus, of the update is negligible. One should keep this in mind when applying the model.

---

[65]To indicate that we focus on an information update with a specific signal, we denote the prior market requirement variance σ_{prior}^2 as $Var(d)$ and the corresponding posterior variance σ'^2 as $Var(d\vert z)$.

Table 5.5. Impact of update for projects with different prior variances ($z = -1.5$)

σ	ξ	$Var(d)$	$Var(d|z)$	$VarRed.$	$V_1(0.5,z)$	$P_1(0.5,z)$	$V_1^I(0.5,z)$
0.1	0.2	0.05	0.02	64%	99.86	86.48	13.38
0.2	0.4	0.20	0.07	64%	96.56	86.48	10.08
0.3	0.6	0.45	0.16	64%	96.07	86.48	9.59
0.4	0.8	0.80	0.29	64%	95.50	86.04	9.46
0.5	1	1.25	0.45	64%	94.70	83.13	11.57
0.6	1.2	1.80	0.65	64%	93.66	83.02	10.64
0.7	1.4	2.45	0.88	64%	92.33	81.03	11.30
0.8	1.6	3.20	1.15	64%	90.74	80.58	10.16
0.9	1.8	4.05	1.46	64%	88.93	79.90	9.04
1	2	5.00	1.80	64%	86.97	78.96	8.01
1.1	2.2	6.05	2.18	64%	84.95	80.97	3.98
1.2	2.4	7.20	2.59	64%	82.82	79.27	3.54
1.3	2.6	8.45	3.04	64%	80.62	80.51	0.11
1.4	2.8	9.80	3.53	64%	78.40	78.33	0.06
1.5	3	11.25	4.05	64%	76.18	76.18	0.00

But note, this observation does not allow any general conclusions about the relation of the prior uncertainty and the value of information. Although we know from Huchzermeier and Loch (2001) and Section 4.2.2 that a decrease of the prior variance increases the posterior project value for a given, fixed uncertainty reduction through the update, this property does not hold for the value of information. As discussed in the analysis of a pure variance update, the payoff differences between the performance states decrease as the prior variances increase since the impact of investments in performance improvements is diminishing. However, the project value based on the prior policy does not necessarily follow this pattern. Depending on the prior uncertainty structure, $P_t(x,z)$ may either in- or decrease as the prior variance increases. Thus, the value of information has no monotone structure.

This characteristic can be illustrated considering the following example. Assume we have identical projects that differ only with respect to the prior market requirement uncertainty. They are all updated at stage $t = 1$ with a signal of $z = -1.5$, which results in every case in a variance reduction by 64%. The results reported in Table 5.5 show that the posterior project value decreases for an increase of the prior variance, while the project value based on the prior policy $P_1(0.5, z)$ does not. Thus, the value of information has, albeit it tends to result in a decrease, the above postulated non-monotone pattern.

So far, we have illustrated and discussed the properties of the developed valuation model in the presence of a specific and known signal value.

The insights obtained from this analysis are, like in the previous chapter, insightful for the analysis of the more complex case, namely the properties in expectation of an unknown signal. This will be discussed next.

5.2.2 Analysis in Expectation of a Signal

Most properties of the model in expectation of a signal, as they are relevant for the valuation of a development project at time $t = 0$, have been derived in Section 4.3. Besides illustrating these properties, this section aims on the one hand at discussing the model characteristics which elude a closed form analysis. On the other hand, we will depict the differences in the optimal managerial policies between the presented basic valuation model and our valuation model, which allows for an uncertainty reduction through a midterm update of the market performance requirements. Based on this analysis, we will identify the key managerial implications for practical applications. We will start with the analysis of the optimal managerial policy.

5.2.2.1 Optimal Managerial Policy

The presented basic valuation model of Huchzermeier and Loch (2001) determines the value of a project given managerial flexibility to respond to the technical uncertainty inherent in the development process. Since the valuation is based on an estimate of the market requirements which determines the expected payoff of the project, this approach is primarily suited for settings where sufficiently precise estimates are feasible. In business environments with high market requirement uncertainty, however, this approach provides an insufficient decision basis since it neglects the possibility to reduce this uncertainty during the development process and to adjust the managerial actions accordingly. A simple revaluation of the project with the basic model at a later stage, when new market information has become available, does not serve the same purpose. Our valuation model, on the other hand, explicitly addresses this issue by incorporating a midterm update of the relevant market information. As the subsequent analysis will show, the former approach leads to suboptimal managerial policies compared to an upfront consideration of an information update as proposed by our model.

When the project is valued at time $t = 0$ under consideration of a later update, the actual value of the signal is unknown. It has therefore to be assumed as random. The determined value $V_t^\tau(x)$ thus corresponds to the optimal project value in expectation of the later update. In particular, the signals are a priori assumed to be symmetrically distributed around the expected market requirement, i.e., $E[f(z)] = E(d)$, which corresponds to

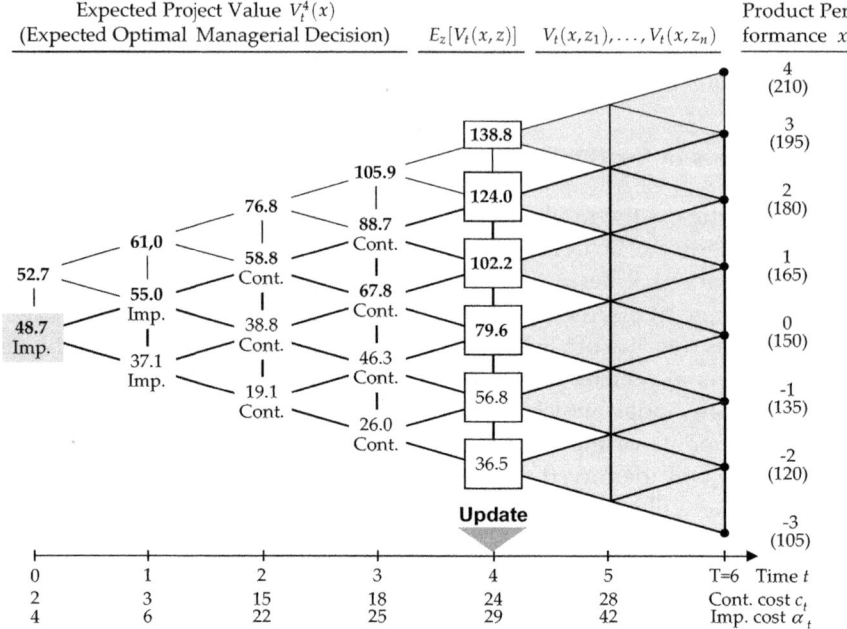

Expected Project Value $V_t^4(x)$
(Expected Optimal Managerial Decision) $E_z[V_t(x,z)]$ $V_t(x,z_1),\ldots,V_t(x,z_n)$ Product Performance x

0	1	2	3	4	5	T=6 Time t
2	3	15	18	24	28	Cont. cost c_t
4	6	22	25	29	42	Imp. cost α_t

Note: See Table 5.7 for the values of $V_t(x,z_i)$; project values of reachable performance states appear in boldface type.

Fig. 5.10. Determination of project value in expectation of update at $\tau = 4$

$E[f(z)] = 0$ in our case.[66] We further assume that the signals are uniformly distributed over $[-\bar{z};\bar{z}]$ and choose, without loss of generality, $\bar{z} = 4$. The selected uniform distribution is, albeit the theoretically most general one, not necessarily the most realistic one. In many practical settings, a symmetrical distribution with mean $E[f(z)] = E(d)$ and flat tails, like the normal distribution for example, may be a more realistic distribution of possible signal values since very low or very high deviations from the forecasted mean will occur less frequently than smaller ones. Although the selected function affects the actual project value, it has no effect on the derived properties. Moreover, in order to reduce the computational effort, we discretize the signal distribution, i.e., $z \in \{-4,-3,\ldots,4\}$. Again, the derived properties are not affected by this simplification.

Based on these assumptions, we can now determine the expected value of the described development project. To ensure the comparability of the so far established numerical results, we choose again a prior standard deviation of $\check{\zeta} = 1.8$ and a performance requirement spread among the cus-

---

[66]See Section 3.3.1.2 for the reasoning behind this assumption.

tomers in the target market of $\sigma = 0.9$. This corresponds to a prior market requirement uncertainty of $Var(d) = 4$, which is reduced through the information update of the market requirement mean to a posterior variance of $Var(d|z) = 1.44$. The estimated expected market requirement (prior mean) remains unchanged, i.e., $\mu = 0$.

If we value the project in expectation of an information update at stage $\tau = 4$ by applying the presented stochastic dynamic program (Eq. 3.105), we obtain in absence of any updating cost an expected project value of $V_0^4(0) = 48.7$ (Fig. 5.10). This information update has an expected value of $V_0^{I,4}(0) = 10.3$. Thus, an update of the market performance requirements at this stage is in expectation valuable for the company if the expenses for the information acquisition do not exceed the benefit from it, i.e., $\frac{\gamma_4}{(1+r)^4} \leq 10.3$.

The expected value of information is, as explained before, a measure of the benefit obtained from an – in expectation of a later market requirement update – adjusted optimal managerial response (Fig. 5.10) compared to the determined counter measures based on the prior project valuation (Fig. 5.3). A comparison of these two specific managerial policies reveals the following: In the first two stages, performance improvements are in both models the optimal managerial action since the corresponding costs are relatively low. In later stages, however, this pattern changes. If management additionally considers in the valuation of the project the possibility of a market requirement uncertainty reduction through a later update, it is optimal to reduce the expenditures for corrective actions until the new information is obtained. This adjustment is in the selected example independent of the considered updating point in time as Table 5.6 shows.

The observed change in the optimal managerial policy is intuitive. Management has with an information update the possibility to learn the true market performance requirements of the customers at stage $t = \tau$. Thus, in expectation of this additional information, it is on the one hand optimal to invest less in improvements until the market requirement uncertainty can be reduced. For the case that the update reveals a lower market requirement as initially estimated, the expenditures would have been useless and would therefore reduce the overall profitability of the project. On the other hand, the number of states where abandonment is the optimal managerial action is also reduced in expectation of a later update since the project could misleadingly be cancelled at any stage before the update although the true market requirement would justify a continuation of the project. Ignoring this effect can have the consequence that the company will have significant sunk costs while missing potential revenues from the project.

In order to explain this pattern of the managerial policy change more formally, we have to study the properties of the two resulting value functions at the updating point in time τ and their implications on the control

Table 5.6. Optimal managerial policies for updates at different stages ($\gamma_\tau = 0$)

Update at	$t = 0$	$t = 1$	$t = 2$	$t = 3$	$t = 4$	$t = 5$
$\tau = 1$	Improve	-	-	-	-	-
$\tau = 2$	Improve	Improve	-	-	-	-
		Improve	-	-	-	-
$\tau = 3$	Improve	Improve	**Continue**	-	-	-
		Improve	**Continue**	-	-	-
			Continue	-	-	-
$\tau = 4$	Improve	Improve	**Continue**	Continue	-	-
		Improve	**Continue**	Continue	-	-
			Continue	Continue	-	-
				Continue	-	-
$\tau = 5$	Improve	Improve	**Continue**	Continue	Continue	-
		Improve	Improve	**Continue**	**Continue**	-
			Improve	**Continue**	**Continue**	-
				Continue	Continue	-
					Continue	-
No update[a]	Improve	Improve	Improve	Continue	Continue	Continue
		Improve	Improve	Improve	Improve	Continue
			Improve	Improve	Improve	Continue
				Improve	Improve	Improve
					Abandon	Continue
						Abandon

[a] Optimal managerial policy based on prior market requirement distribution.

Note: The managerial policies that differ from the ones based on the prior market requirement distribution appear in boldface type.

limits of the optimal managerial policy. We know from Huchzermeier and Loch (2001) that the following holds:

Lemma 5.1. *Suppose there are two value functions $V_{t+1}(x)$ and $\overline{V}_{t+1}(x)$ that are both convex-concave increasing in x and fulfill the following characteristic:*

$$\overline{V}_{t+1}(x) - \overline{V}_{t+1}(x-1) \geq V_{t+1}(x) - V_{t+1}(x-1) \quad \forall x, \tag{5.1}$$

then $V_t(x)$ and $\overline{V}_t(x)$ are as well convex-concave increasing and fulfill condition Eq. 5.1.

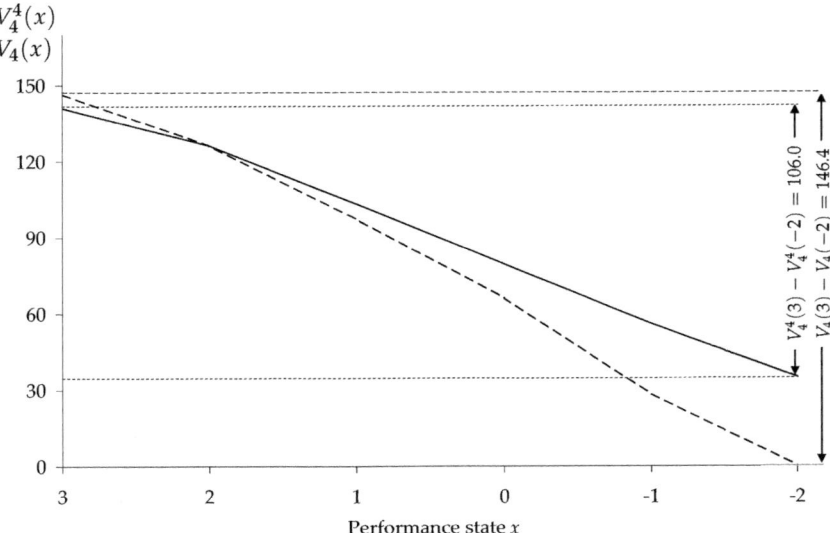

Fig. 5.11. Variability of (expected) project value function

Proof. See Huchzermeier and Loch (2001) [Lemma 3].

In other words, if the described properties hold for a value function at stage $t + 1$, they also hold for the resulting value function of the stochastic dynamic program at stage t. Based on the considerations of Appendix B, this lemma can in principle be applied to our valuation model. It allows us to infer from the characteristics of the value functions at stage $t = \tau$ backwards on the control limits of the optimal managerial policies at stages $t < \tau$, provided that the conditions above are satisfied. Denote for the following considerations with the upper bar all limits and results corresponding to the value function with the larger increments.

If the variability of project increases, we know by the definition of the (optimal) managerial policy control limits (cf. Eq. B.3)[67] that the upper control limit increases, i.e., $\overline{L}_u \geq L_u$, while the lower one decreases, i.e., $\overline{L}_m \leq L_m$, . In other words, $\overline{V}_t(x)$ has a larger improvement region than $V_t(x)$ since $\overline{V}_{t+1}(x)$ is steeper by definition (Eq. 5.1). The abandonment region on the other hand is larger for $\overline{V}_t(x)$, i.e., $\overline{L}_d \geq L_d$, if $\overline{V}_{t+1}(x) \leq V_{t+1}(x)$ for $x \leq \overline{L}_d$.

Applied to our case, we can observe for the selected example that $V_t^\tau(x)$ is at stage $t = \tau$ convex-concave increasing in x (see Fig. 5.11) while $V_t(x)$ is

---

[67]See Appendix B as well as Section 4.1 for details.

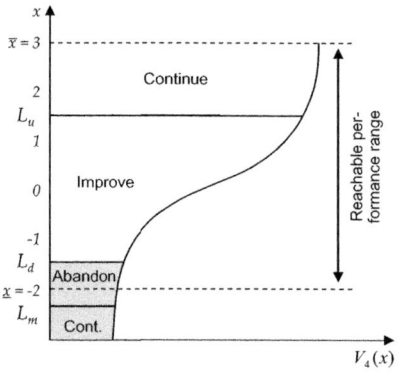 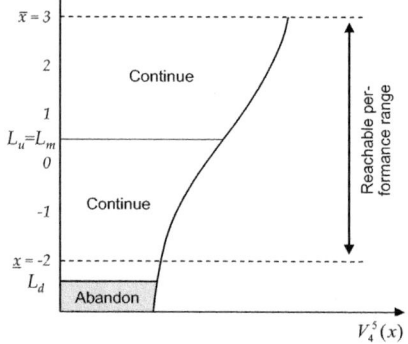

(a) Prior project value function (basic model)

(b) Expected project value function (information updating valuation model)

Fig. 5.12. Change of control limits for optimal managerial policy

convex-concave increasing by definition.[68] In addition, the expected value function has at stage $t = \tau$ a lower variability than the prior value function, i.e., $V_t^\tau(\bar{x}) - V_t^\tau(\underline{x}) < V_t(\bar{x}) - V_t(\underline{x})$. This implies that $V_t(x)$ has larger increments than $V_t^\tau(x)$ and Lemma 5.1 can be applied. Thus, the improvement region of the expected project value function for any stage $t < \tau$ is smaller compared to the project value function determined on the basis of the prior distribution, i.e., in expectation of a later market requirement update it is optimal to spend less in improvements until the additional information is acquired.

The reduction of the improvement region also becomes apparent from Fig. 5.12 which represents schematically the control limits of the optimal policy for the considered examples at stage $t = 4$. The left part shows the value function of the basic model (without an updating possibility) while the right part depicts the value function of our model in expectation of a later updating possibility. We can observe that the reduced variability of the project function due to the expected update at $\tau = 5$ does not justify improvements at stage $t = 4$ anymore (see also Table 5.6). Since we consider only discrete performance states, the control limits L_u and L_m of the improvement region even collapse (Fig. 5.12 b). Simultaneously, the lower control limit L_d falls below the lowest reachable performance state $\underline{x} = -2$

---

[68]Note that although the expected value function fulfills in most cases the convex-concavity requirement due to the a priori assumed symmetrical distribution of the expected signals, this property does not hold in general. The reason is that even if $V_t(x, z)$ is convex-concave increasing in x for all $z \in \mathbf{Z}$, $V_t^\tau(x) = E_z\left[V_t(x, z)\right] (t = \tau)$ does not necessarily have to be convex-concave.

so that abandonment is not an optimal managerial action for any performance state of this stage if the market performance requirements can be updated at the next stage.

This effect increases with an increase of the uncertainty about the future development of the market performance requirements. The variability increase diminishes the payoff differences over the reachable performance states (i.e. the payoff function becomes flatter) and hence, reduces the option value of managerial flexibility. Thus, in the extreme case, continuation is the optimal managerial action until the updating point in time. Thereafter, however, the market requirement uncertainty is reduced and the updated payoff function becomes much steeper. Thus, the optimal managerial policy of the remaining stages changes significantly: Improvement becomes valuable in the performance states where the payoff function is at its steepest and abandonment may be worthwhile in the lower performance states. But note, if the expected signal range is relatively small or extreme signals are regarded as less likely, the expected value function deviates only slightly from the one based on the prior information, i.e., the improvement region is only slightly reduced, and the policy change is less dramatic.

Note that if $\underline{x} \geq L_d$ for all $\gamma_\tau (1 + r)^{-(\tau-t)} \leq V_t^{I,\tau}(x)$ (L_m may be anywhere with respect to $\underline{x}$), the reduction of the improvement region compared to the one of the basic valuation model is independent of the updating cost. The reason is that in this case the expenses for the update result only in a parallel shift of the value function, but do not change its shape within the reachable performance range of the project. The abandonment region, on the other hand, will be affected by the updating cost. However, the increase of L_d through an increase of the updating cost may be irrelevant for the project valuation. Like in the considered example, this is the case if the lower control limit remains below the lowest reachable performance state of the project even for an update at the highest reasonable updating cost, i.e., $\frac{\gamma_\tau}{(1+r)^\tau} = V_0^{I,\tau}(0)$.

These considerations confirm the intuition above that the managerial policy changes in the presence of market requirement uncertainty with a later updating possibility compared to the project valuation based on the prior market performance requirement information. A simple revaluation of the project with the basic decision model at each review point, as proposed by Santiago and Bifano (2005), based on the best available information does not serve the same purpose. Since a later update of the market performance requirements represents an option value from today's perspective, this possibility has to be considered upfront in the determination of the optimal managerial response to the technical development uncertainty. The "passive" approach of the basic model to wait and see until new information becomes available ignores this option and therefore leads to an inferior managerial policy, e.g., unnecessary investments in performance improvements and precipitate abandonments.

This becomes particularly apparent under the generally applicable assumption of increasing improvement costs. Under such a cost structure, the basic valuation model generally leads to early improvements when the costs for counter measures are still low (Huchzermeier and Loch 2001). As this analysis has shown, however, this strategy is in the presence of market requirement uncertainty no longer optimal. Instead, it is more valuable to invest upfront less in improvements and respond afterwards with high precision to the new information. Thus, a flexible development process as well as a sophisticated product design that allows for late design changes at relatively low costs facilitates this strategy. The corresponding impact of the underlying cost structure will therefore be studied among other properties of our valuation model in the next section.

5.2.2.2 Cost Structure Changes

The structure of the improvement costs has, as the discussion in Section 4.1 has shown, a decisive impact on the project value since it allows management to respond to the technical development uncertainty. In the presence of high market requirement uncertainty with the possibility of a later update, this effect is even increased. The analysis of the managerial policy in the previous section has shown that, in expectation of a later uncertainty reduction through additional information, it is generally optimal to reduce the efforts for corrective actions prior to the update compared to a project setting without this updating possibility. The reason for this pattern change is that without any information in which direction the performance requirements of the customers might drift, high expenditures for performance improvements may be unnecessary and thus, reduce the overall profitability. Hence, improvements are only conducted in expectation of a later update in those development stages, where the costs of these actions are so low that the overall expected benefit of an upfront adjustment exceeds the cases where these actions are superfluous.

After the acquisition of additional information, however, management can precisely respond to the updated market performance requirements by selecting the optimal managerial action in dependence of the project's technical performance that has been realized until this point in time. The room for maneuver hereby significantly depends on the improvement costs of the remaining development stages. Although the information update reduces the variance of the market requirement distribution, which results in a steeper convex-concave payoff function and hence, in a larger improvement region, the improvement costs at the later stages are generally significantly

Table 5.7. Expected project value for update at $\tau = 4$ ($\alpha_4 = 29$)

$V_0^4(0)^a$	z=-4			$\cdots$	z=0			$\cdots$	z=4		
	$V_4(x,-4)$	$V_5(x,-4)$	$\Pi(x,-4)$		$V_4(x,0)$	$V_5(x,0)$	$\Pi(x,0)$		$V_4(x,4)$	$V_5(x,4)$	$\Pi(x,4)$
48.7	163.2 (C)	198.4 (C)	240.0		146.9 (C)	192.3 (C)	238.5		26.9 (I)	63.7 (I)	104.1
	163.0 (C)	198.4 (C)	240.0		117.9 (C)	170.1 (C)	228.5		0 (A)	0 (A)	38.1
	161.0 (C)	198.0 (C)	239.9		87.7 (C)	128.1 (I)	191.4		0 (A)	0 (A)	8.0
	150.9 (C)	194.2 (C)	239.1		43.7 (C)	76.9 (I)	120.0		0 (A)	0 (A)	0.9
	122.9 (C)	176.7 (C)	232.0		0 (A)	9.5 (I)	48.6		0 (A)	0 (A)	0.1
		134.7 (I)	201.9			0 (A)	11.5			0 (A)	0.0
			135.9				1.5				0.0

Stage t	4	5	6		4	5	6		4	5	6
Cont. cost c_t	24	28			24	28			24	28	
Imp. cost α_t	29	42			29	42			29	42	

[a] Determined based on a signal range of $z \in \{-4, -3, \ldots, 4\}$ with a discretized uniform distribution, i.e., $f(z_i) = 1/9$, and in absence of any updating cost, i.e., $\gamma_t = 0$.

Note: The optimal managerial decision, i.e., continue, improve, or abandon, in each state is provided in brackets.

higher compared to the one of earlier periods. Thus, the absolute improvement costs of the remaining development stages determine not only the project value, but also the benefit of information updating.

This effect can be illustrated considering the example of the previous section with an update of the market performance requirement distribution at stage $\tau = 4$ that results in an expected project value of $V_0^4(0) = 48.7$ for the given cost structure. Due to the high improvement costs at stages $t = 4$ and $t = 5$, counter measures to improve the performance of the project are – regardless of the observed signal – only worthwhile for performance states around the updated market requirement mean where the payoff function is at its steepest (see Table 5.7). For the performance stages in the flatter regions of the payoff function, however, the higher payoff does not justify the high improvement costs of the remaining periods. Thus, management only has the possibility left to abandon the project if even the continuation costs are not earned anymore.

However, a relatively small reduction of the improvement cost at stage $t = 4$ from $\alpha_4 = 29$ to $\alpha_4' = 25$ increases the improvement range in that period for almost all possible signal values (see Table 5.8). Solely for a signal of $z = 4$, which lies beyond the reachable range of the project, this improvement cost reduction does not change the managerial policy. But over all possible signal values, the expected project value increases from $V_0^4(0) = 48.7$ to $V_0^4(0) = 49.9$.[69] This implies that it is beneficial at the beginning of the project to invest up to this increase, i.e., $\delta_1 = 1.2$, in a more flexible design that reduces the improvement cost at stage $t = 4$ by $\delta_4 = 4$ to $\alpha_4' = 25$. Although the maximum upfront investment cost is significantly lower than the corresponding time value of money of the improvement cost reduction, i.e., $\delta_1 = 1.2 < \delta_4(1 + r)^{-4} = 3.2$, this result still justifies managerial actions that lead to such a cost structure change. Besides the general insight that early design changes induce significantly lower expenses than later ones, one has to consider that, additionally, a more flexible design generally reduces the costs for corrective actions of more than one development stage.

One possibility to reduce the costs of later performance improvements for the considered example of a power supply measurement card is to implement some of the test routines directly on the card. This requires, on the one hand, higher expenditures upfront for the design of the additional hardware components and the development of the corresponding test routines (software). On the other hand, the on-board self-testing feature allows faster prototyping cycles and reduces the testing costs for modifications of the card

---

[69]Without the consideration of a later updating possibility, the same cost structure change would only result in an increase from $V_0(0) = 48.3$ to $V_0(0) = 49.1$. Thus, the valuation of the project based on the basic model systematically underestimates the true value of such a strategy.

Table 5.8. Expected project value for update at $\tau = 4$ ($\alpha_4 = 25$)

$V_0^4(0)$	$z=-4$			$\cdots$	$z=0$			$\cdots$	$z=4$		
	$V_4(x,-4)$	$V_5(x,-4)$	$\Pi(x,-4)$		$V_4(x,0)$	$V_5(x,0)$	$\Pi(x,0)$		$V_4(x,4)$	$V_5(x,4)$	$\Pi(x,4)$
49.9	163.2 (C)	198.4 (C)	240.0		146.9 (C)	192.3 (C)	238.5		26.9 (I)	63.7 (I)	104.1
	163.0 (C)	198.4 (C)	240.0		121.9 (I)	170.1 (C)	228.5		0 (A)	0 (A)	38.1
	161.0 (C)	198.0 (C)	239.9		91.7 (I)	128.1 (I)	191.4		0 (A)	0 (A)	8.0
	150.9 (C)	194.2 (C)	239.1		47.7 (I)	76.9 (I)	120.0		0 (A)	0 (A)	0.9
	125.9 (I)	176.7 (C)	232.0		0 (A)	9.5 (I)	48.6		0 (A)	0 (A)	0.1
		134.7 (I)	201.9			0 (A)	11.5			0 (A)	0.0
			135.9				1.5				0.0

Stage t	4	5	6		4	5	6		4	5	6
Cont. cost c_t	24	28			24	28			24	28	
Imp. cost α_t	**25**	42			**25**	42			**25**	42	

[a] Determined based on a signal range of $z \in \{-4, -3, \ldots, 4\}$ with a discretized uniform distribution, i.e., $f(z_i) = 1/9$, and in absence of any updating cost, i.e., $\gamma_t = 0$.

Note: The optimal managerial decision, i.e., continue, improve, or abandon, in each state is provided in brackets. The ones that have changed through the cost reduction appear in boldface type.

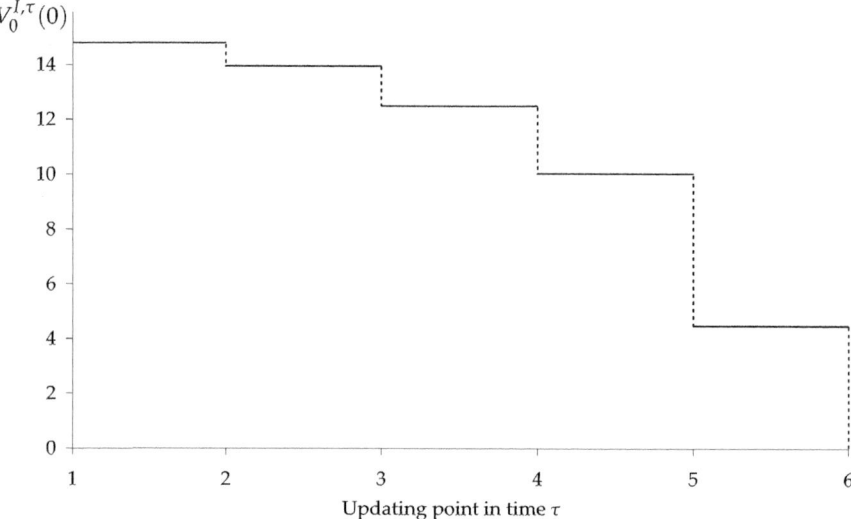

Fig. 5.13. Expected value of information as a function of updating point in time

since it does not always require an adjustment of the external test equipment setup. Thus, the broader design upfront causes higher investment costs but simultaneously reduces the improvement as well as the continuation costs of the project in the late design (stage $t = 4$) and the qualification phase (stage $t = 5$).

These considerations show that cost reductions through a more flexible design at the expense of higher upfront investment costs can be worthwhile if management has the possibility to reduce the initial market requirement uncertainty through a later information update. The derived properties when and under which conditions such an update is in expectation valuable will be illustrated next.

5.2.2.3 Optimal Updating Point in Time

The formal analysis of the model in Section 4.3.2 revealed that the expected value of information $V_t^{I,\tau}(t)$ increases if the updating point in time decreases (cf. Proposition 4.11). In other words, the earlier the update is conducted, the higher the expected value of information. Fig. 5.13 shows the expected value of information for state $x = 0$ at stage $t = 0$, i.e., $V_0^{I,\tau}(0)$ over different updating points in time τ for the chosen example of the previous section with a prior market requirement uncertainty of $Var(d) = 4$ ($\xi = 1.8$ and $\sigma = 0.9$). As stated before, the reason for this decrease of the expected value of information for a postponement of the update is that the earlier the in-

Table 5.9. Expected value of information for different updating points in time

σ	ζ	$Var(d)$	$Var(d\vert z)$	Var Red.	$V_0^{I,1}(0)$	$V_0^{I,2}(0)$	$V_0^{I,3}(0)$	$V_0^{I,4}(0)$	$V_0^{I,5}(0)$
0.1	1.998	4	0.02	99.5%	27.52	25.58	24.29	20.35	12.61
0.2	1.990	4	0.08	98.0%	27.42	25.49	24.16	20.19	12.45
0.3	1.977	4	0.18	95.6%	26.94	25.05	23.68	19.70	11.99
0.4	1.960	4	0.31	92.2%	25.60	23.80	22.42	18.48	11.04
0.5	1.937	4	0.48	87.9%	23.68	22.03	20.64	16.86	10.04
0.6	1.908	4	0.69	82.8%	21.50	20.04	18.63	15.04	8.87
0.7	1.874	4	0.92	77.0%	19.26	17.97	16.54	13.30	7.50
0.8	1.833	4	1.18	70.6%	16.95	15.87	14.42	11.61	5.98
0.9	1.786	4	1.46	63.6%	14.66	13.83	12.38	9.93	4.38
1	1.732	4	1.75	56.2%	12.44	11.87	10.59	8.18	3.25
1.1	1.670	4	2.05	48.6%	10.26	9.96	8.92	6.42	2.48
1.2	1.600	4	2.36	41.0%	8.11	8.07	7.26	4.98	1.73
1.3	1.520	4	2.67	33.3%	6.41	6.41	5.68	3.88	1.07
1.4	1.428	4	2.96	26.0%	5.05	5.05	4.28	2.88	0.53
1.5	1.323	4	3.23	19.1%	3.80	3.80	3.10	2.03	0.25

formation update is conducted, the more development stages and review points remain where the managerial policy can be optimally adjusted to the acquired additional market requirement information. Hence, the value of information is in expectation of an update more valuable at earlier development stages compared to later ones. Recall that this does, however, not imply that the ideal updating moment is the earliest point in time when the expected value of information exceeds the updating cost γ_t. It is only the necessary condition for the update while the highest expected project value under consideration of the updating cost structure that fulfills this condition determines the ideal updating point in time.

What also becomes apparent from Fig. 5.13 is the fact that the later the update is postponed, e.g., by one period, the higher the reduction of the expected value of information gets. This effect results from the underlying cost structure. The high improvement costs at late stages of the development process limit the benefit of managerial counter measures to respond to the information gain. Thus, the discrepancy between the optimally adjusted managerial policy and the one based on the prior market requirement distribution is lower compared to an earlier update.

The just described pattern holds, as shown before, in general. On the one hand, the expected value of information increases for any given market requirement uncertainty structure as the updating point in time τ decreases. On the other hand, we know that an increase of the uncertainty reduction through the update leads to an increase of the expected value of information

(Proposition 4.10). Table 5.9 reports the expected value of information for a common prior variance of the market requirement distribution composed of the prior standard deviation of the mean (ξ) and the requirement spread among the customers in the target market (σ) that result in different posterior variances and hence, in different uncertainty reductions. For reasons of consistency to the previous sections and without loss of generality, we pick again $Var(d) = 4$.

We can observe that the expected value of information increases with the increase of the uncertainty reduction through the update. The intuition behind this property is the following: The posterior variance of the market requirement distribution decreases as the uncertainty reduction increases. A lower variance, however, results in a steeper expected payoff function and thus, increases the improvement region for any observable signal. The discrepancy between the optimal adjusted managerial policy and the one based on the prior distribution increases. Since this property holds for any signal, this effect is additive and the expected value of information has the described property. Simultaneously and as proposed above, it is apparent that the expected value of the update decreases for any specific market requirement variance structure if the information update is postponed by one period (Table 5.9).

In absence of any updating cost, i.e., $\gamma_t = 0$ for all $t = 1, \ldots, T - 1$, the former property of an increase in the expected value of information for a postponement of the updating point in time τ, holds, for the just described reasons, also for the expected project value $V_t^\tau(x)$ (Proposition 4.9). Table 5.10 shows this increase of the expected project value over various uncertainty structures for the above described project setting. Note that in contrast to the expected value of information, the expected project value does not necessarily decrease if the uncertainty reduction increases. Based on its definition, $V_t^\tau(x)$ determines the value of the development project in expectation of a later update with an a priori symmetrically distributed signal z. It therefore simultaneously considers mean and variance effects. But based on the consideration in Section 5.2.1.2, it is obvious that this does in expectation not always result in a monotonous pattern of the project value if the variance is varied over the feasible range. This effect especially occurs for late updating points in time (e.g., for an update at stage $\tau = 4$ or $\tau = 5$ in the considered example) when management has only very few stages left to respond to the information gain with appropriate counter measures. For certain updating points in time, a lower posterior variance may therefore result in a higher expected project value. This is, however, no contradiction to the results derived above since the uncertainty structure of the project is given and thus, an exogenous variable.

This leaves us with the remaining issue regarding the a priori optimal updating point in time. We assumed, in compliance with other models of

Table 5.10. Expected project value for different updating points in time ($\gamma_t = 0$)

σ	ξ	$Var(d)$	$Var(d\|z)$	Var Red.	$V_0^1(0)$	$V_0^2(0)$	$V_0^3(0)$	$V_0^4(0)$	$V_0^5(0)$
0.1	1.998	4	0.02	99.5%	59.12	57.18	55.89	51.95	44.21
0.2	1.990	4	0.08	98.0%	59.32	57.39	56.06	52.09	44.35
0.3	1.977	4	0.18	95.6%	59.25	57.35	55.98	52.01	44.29
0.4	1.960	4	0.31	92.2%	58.57	56.78	55.39	51.45	44.02
0.5	1.937	4	0.48	87.9%	57.55	55.91	54.52	50.74	43.92
0.6	1.908	4	0.69	82.8%	56.44	54.98	53.57	49.98	43.81
0.7	1.874	4	0.92	77.0%	55.39	54.10	52.67	49.43	43.63
0.8	1.833	4	1.18	70.6%	54.36	53.28	51.82	49.01	43.39
0.9	1.786	4	1.46	63.6%	53.40	52.56	51.11	48.66	43.11
1	1.732	4	1.75	56.2%	52.51	51.94	50.66	48.25	43.32
1.1	1.670	4	2.05	48.6%	51.64	51.35	50.31	47.81	43.86
1.2	1.600	4	2.36	41.0%	50.76	50.72	49.91	47.64	44.38
1.3	1.520	4	2.67	33.3%	50.25	50.25	49.52	47.72	44.91
1.4	1.428	4	2.96	26.0%	49.97	49.97	49.20	47.80	45.45
1.5	1.323	4	3.23	19.1%	49.68	49.68	48.98	47.91	46.13

this type, that management has the possibility to obtain the additional information to update its initial estimates at anytime during the development process; solely the costs for the signal vary over time. As explained in detail in Section 3.3.1, it is reasonable to assume in the context of new product development projects that the cost of information acquisition decreases over time. While the true performance requirements of the customers may be easily and inexpensively obtainable close to the launch of the product, it requires significantly higher resources to acquire the same information at earlier development periods provided that it is actually feasible.

For the described development project of the power supply device, this issue turns out to be as follows. For most development projects in the semiconductor industry, it is generally impossible to obtain a precise estimate of the customers' performance requirements at the start of a project. Since the development process of measurement devices, like the one considered here, is closely linked to the development of its applications, e.g., memory chips, the customers do not know themselves precisely two or three years prior to the launch of their own products which performance they will really need. Thus, an update of the prior market requirement estimates at the definition phase of the project is therefore infeasible and the updating costs have to be set accordingly. Just for ease of exposition in the subsequent figures, we pick a sufficiently large number of the updating cost of stage $\tau = 1$ that fulfill the condition $\frac{\gamma_1}{(1+r)} \gg V_0^{I,\tau}(0)$, e.g., $\gamma_1 = 20$ instead of setting it as infinity.

Table 5.11. Estimated updating costs

Time t	1	2	3	4	5
Updating cost γ_t	20	18	6	3	1

Once the customers have defined their own product specifications, the performance requirements become more precise. By tightly linking the development process with the R&D activities of key customers, it is generally possible to accelerate the information gain about the performance requirements by providing them with appropriate prototypes earlier as initially planned. This approach requires, however, significant additional resources in dependence of the desired acceleration. For the system-level design phase, we assume that such an effort would require extra resources that equate the development cost of the subsequent development phase in addition to the current continuation cost of $c_3 = 3$, i.e., $\gamma_2 = 18$.

In addition to these efforts, management can derive an update of the performance requirements also from similar products that are currently launched or announced. These market studies are generally at the second half of the development process a fairly precise and less expensive source of information. Thus, from the design phase on, this information acquisition approach can gradually be used instead. The utilization of this information source requires, however, an additional assumption in order to comply with the presented Bayesian updating formulation. Since these data are generated from similar, but not the same product under consideration, we have to assume in this particular case that the sample variance σ^2 reflects uncertainty about both the variance about the performance requirements between the different customers as well as the use of data from similar, but not the same product as an information source to update the initial estimates.[70] In our example, this assumption has, however, no implications for the sample variance since the latter variance element is in comparison to the former relatively small and thus, negligible. The updating costs for the remaining development stages have been estimated as summarized in Table 5.11.

The comparison of this cost structure with the previously determined expected value of information for the different updating points in time shows that in the first two periods an update of the market performance requirements is not valuable (see Fig. 5.14); the updating costs are significantly higher than the value that can be expected from an update at either one of the first two review points. But from stage $\bar{\tau} = 3$ on, the expected value of

---

[70]Iyer and Bergen (1997), for example, make a similar assumption for the uncertainty reduction in their Quick Response model.

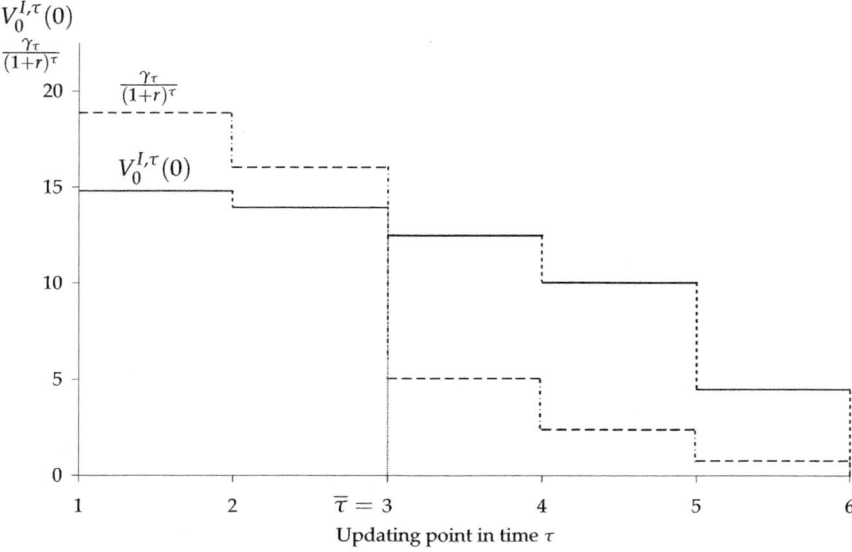

Fig. 5.14. Updating limit

information exceeds these costs and the acquisition of additional information yields an positive expected return on this investment for the company.

The analysis of the expected project value under consideration of this estimated updating cost structure shows that the ideal updating point in time is $\tau^* = 4$ (Table 5.12), which satisfies the necessary condition above, i.e., $\tau^* > \bar{\tau}$. It can be observed, however, that the difference between the expected value of information for an update at stage $\tau = 3$ and one that is conducted at $\tau = 4$ is very small which indicates a high sensitivity of the ideal updating point in time to the estimated updating cost structure. Due to this diminishing difference, one should consider to conduct the update (contrary to this result) already at stage $\tau = 3$, which allows management to obtain the updated market requirements earlier and hence, to respond accordingly to it. Whether the update is conducted at stage $\tau = 3$ or at $\tau = 4$, has, however, no impact on the optimal managerial policy in any state prior to the update (see Table 5.6).

The result of the upfront analysis of the ideal updating point in time should therefore be regarded rather as a (rough) indicator at which phase of the development process the update should be conducted (or at least from which period on it will be beneficial to conduct it) than as a strict instruction. Even if not the optimal updating point in time is found a priori, this approach provides management with a basis for such a decision. In addition, the obtained insights from such an analysis allow management to adjust its optimal managerial response to this information and, as the discussion of

Table 5.12. Expected project value

Updating point in time τ	1	2	3	4	5
Expected project value $V_0^\tau(0)$	34.6	36.6	46.1	46.3	42.4

the previous section revealed, avoid too high investments in performance improvements at early development stages. The developed valuation framework thus compensates for the shortcomings of the basic valuation model.

5.3 Summary

In this chapter, we have conducted a comparative static analysis based on a real-life investment project faced by a manufacturer of semiconductor test equipment. The data obtained from this on-going development project ensured realistic dimensions of the different model parameters for the conducted analysis and simultaneously demonstrated the practical applicability of the model. Besides the illustration of the properties that have been derived in closed form in Chapter 4, we studied the impact of a general update, i.e., an update of both mean and variance, on the project value. It turned out that only a moderate mean increase can be compensated by a simultaneous variance reduction accordingly. If the mean shift is too large, either the project target has to be redefined or different counter measures have to be considered.

The comparison of the optimal managerial policy derived with our model and the one proposed on the basis of the valuation framework developed by Huchzermeier and Loch revealed that a later updating possibility has to be explicitly considered upfront. In expectation of a later update of the market performance requirements, it is generally optimal to exercise less options, i.e., to improve or abandon the project, prior to the update. The simple strategy underlying the basic model of revising the managerial policy when new information becomes available does not adequately incorporate the updating possibility in the determination of the managerial policy. Thus, this approach may lead, on the one hand, to unnecessary investments in improvement while, on the other hand, to a precipitate abandonment of the project, albeit these actions are not optimal under consideration of the new information.

Moreover, the numerical study revealed insights on the impact of an improvement cost reduction at a later stage at the expense of higher upfront investment costs. Up to the identified threshold value for this specific project, an investment in a more flexible design increases the project value. Finally, the analysis of the in expectation optimal updating point in time allowed to determine the sensitivity of the derived result on the underlying model parameters.

6

Conclusion

We have developed a decision model that allows to valuate the impact of
information updating in new product development projects. This has been
accomplished by integrating a general updating mechanism for market requirement information during the development process into a real options
framework. Our approach allows to explicitly address two sources of uncertainty: firstly, technical uncertainty stemming from the performance variability of the project and secondly, market uncertainty arising from performance requirement variability of the customers. While management can respond to the former uncertainty through appropriate counter measures, the
latter can be reduced by updating the initial market requirement estimates
with information obtained from additional market studies, for example.

More precisely, in the presented real options model of Huchzermeier
and Loch (2001), which served as the basis for our framework, management
has at each review point of the development project the flexibility to react contingent on the reached technical performance state by choosing one
of the following actions: continuing the project, improving it through investments in additional resources, or abandoning it if the expected value of
the other two actions does not justify the corresponding expenses. Although
this framework values managerial flexibility in the presence of development
uncertainty and thus, addresses the shortcomings of the traditional NPV
methods, obtaining new information is regarded as a passive consequence
of the project progression. The (active) acquisition of additional information
is, as in most other real options frameworks, not explicitly considered.

However, especially NPD projects are exposed to numerous sources of
uncertainty forcing management to trade off between either an early, but inexpensive decision based on uncertain prior information or a postponement
until more information becomes available at the expense of higher costs. Our
approach to integrate an information updating mechanism into the above
mentioned decision framework allows to address this issue by evaluating

C. Artmann, *The Value of Information Updating in New Product,*
DOI: 10.1007/978-3-540-93833-0_6, © Springer-Verlag Berlin Heidelberg 2009

its impact on the project value and hence, on the optimal managerial policy. Moreover, the value of additional information can explicitly be determined with this framework and the provided multidimensional model formulation ensures the practical applicability to NPD projects. The formal analysis of the model as well as the numerical study revealed insightful results that provide valuable contributions to the current research in this field. They are summarized below.

6.1 Contribution to Current Research

The typical investment decision process for a NPD project starts with its valuation based on the available information at that time, derived from forecasts, similar or past projects, or expert judgment. If this initial assessment yields a positive option value, the project will be started. Although real options frameworks, like the model of Huchzermeier and Loch (2001), take the managerial flexibility into account to respond to contingencies during the development process, they generally ignore the incorporation of additional information obtained at later stages. As our analysis has shown, however, such an updating possibility to reduce market uncertainty has substantial value as it affects the optimal execution of options. The strategy proposed by Santiago and Bifano (2005) of revising the initially determined optimal policy at each stage with the at that point in time available information does not adequately address this aspect. It corresponds to the above mentioned passive strategy, underlying most real options frameworks, of waiting until new information becomes available, which generally leads to an early execution of options although this strategy is suboptimal in expectation of a later update. Thus, the managerial policy determined on the basis of the former model leads to wrong decisions prior to the update and hence, to a lower project value. This result is therefore somewhat contrary to the result of the above mentioned authors.

With the idea of integrating an information updating mechanism into a real options framework, the model is in line with some recently published frameworks that combine Bayesian analysis and real options concepts in a similar way (Miller and Park 2005; Armstrong et al. 2005). In contrast to our approach, these models primarily focus on the value of learning. In addition, they are relatively simple in their structure and do not address the specific characteristics of NPD projects. Our model formulation, on the contrary, takes these characteristics into account. Besides the common case of a sole mean update, it also allows for an update of the variance, and both, the mean and the variance. The consideration of these cases enhances the practical applicability of the valuation framework. Furthermore, the derived Bayesian updating mechanism builds upon the conjugate relationship between the prior and the likelihood distribution, which is a more general and robust concept compared to the frequently (and often unreflectively) assumed bivariate normal relationship.

It should be noted, however, that Bayesian analysis is just one possibility to model information updates for uncertainty reduction in decision problems, which also has some limitations. Compared to other approaches, like time series analysis or similar frequentist approaches for example, Bayesian analysis relies on subjective estimates. Thus, a frequently criticized issue is the difficulty to determine an appropriate prior distribution that adequately reflects the available knowledge about the unknown parameter(s). In addition, Bayesian analysis requires that the unknown quantities can be described with probability distributions. Evidently, all inferences which are made from the combination of prior information with later observed data depend on the initial subjective beliefs. For example, the interpretation of the same available forecasts about market performance requirements by different experts could result in different prior estimates and hence, yield different conclusions and decisions (Miller and Park 2005). Nevertheless, such difficulties occur in any decision problem under uncertainty where assumptions have to be made due to the lack of sufficient information. On the contrary, the ability to formally combine prior information with other sources of information is one of the key advantages of the Bayesian approach over the other, frequently applied updating methods. Moreover, it corresponds to the natural decision making process in most companies of having initially subjective estimates about uncertain outcomes of the project which are later revised and updated when new information becomes available. As the numerical study has demonstrated, it is therefore well suited for decision making in development projects with high market requirement uncertainty, which applies to the largest fraction of such projects.

All comparable decision models in the field of operations management that use a Bayesian method to update information, like the above cited Bayesian real options models or the Quick Response models in the area of supply chain management for example, are – to the best of our knowledge – one-dimensional. As Loch and Terwiesch (2005) claim, however, NPD projects are generally more complex so that the information exchange rarely exhibits the one-dimensional, ordered structure. To overcome this deficiency of the existing Bayesian updating models, we have also provided a multi-dimensional model formulation that allows to consider multiple product performance parameters. As it is sufficient for the derivation of the model properties to consider solely the more intuitive one-dimensional case, we maintained this perspective in the numerical study for ease of representation. Despite the fact that the Bayesian updating formulation as well as the entire decision model becomes more complex in the multidimensional case, it is still manageable and applicable to real-life projects, especially if only a well specified range of dimensions is considered. A focus on the two or three most important and decisive parameters should be sufficient for most projects.

A comparable model for decision making in NPD projects based on preliminary information has only been provided by Loch and Terwiesch (2005) so far. They describe information structures in form of a sigma field that is refined over time. This approach allows to consider multiple dimensions of actions and outcomes. Although this model is very insightful and the first of its kind with a clear focus on NPD decision characteristics, it is not suited, as the authors stress themselves, to quantitatively support managerial decisions. With this respect, our model is the first that provides a decision and valuation framework for an information update on multiple parameters in a Bayesian manner and hence, overcomes the deficiencies of the existing frameworks regarding their general applicability to NPD project settings.

With this approach, we are not only able to quantitatively determine the optimal managerial policy, but also to determine the value of such an information update. The value of additional information critically depends on the decision maker's managerial flexibility (Merkhofer 1977). Only if the firm has the possibility to react to new insights, e.g., changed performance requirements of key customers, the additional information is of substantial value. Without the possibility to respond, the value of information is diminishing. While this coherency is well-known and its impact on the success of NPD projects has been empirically studied and validated (cf. e.g., Thomke 1997; Upton 1995), hardly any measures exist to determine the exact value of such flexibility in the development process. With the developed framework, management can evaluate this flexibility and determine the threshold up to which investments in design flexibility for less expensive design changes at later stages are valuable.

Thus, the developed decision model contributes to close this gap and reduces the frequently complained lack of quantitative models. With our framework, we address the needs identified in empirical studies for such models as well as the criticism raised by several authors in the field of operations management who claim the missing focus on product innovation and NPD project issues of the currently existing models.

6.2 Managerial Implications

The above discussed findings have clear managerial implications. In the following, we will summarize these insights in a set of managerial principles and discuss how they can be used by firms in order to improve decision making in their NPD projects.

Principle 1: Take an information updating possibility explicitly into account when valuing the project a priori.

The possibility of a later update allows management to reduce market requirement uncertainty with information obtained from additional market studies, for example. Based on the updated beliefs about the customer requirements, management can revalue the project and adjust the previously determined optimal managerial actions accordingly. However, the later possibility of responding to additional information affects the optimal managerial policy over the entire development process and hence, the optimal actions prior to the update. Thus, an information updating possibility at later development stages must always be explicitly taken into account when valuing the project. It represents from an option's point of view a value in itself (Armstrong et al. 2005; Miller and Park 2005) that, if neglected, leads to wrong managerial decisions and thus, to a lower overall project value.

Principle 2: Exercise less managerial options prior to the additional information gain.

This principle follows directly from the first one. The valuation of the project with the presented basic real options framework leads to investments in performance improvements early in the development project when these actions are still inexpensive compared to later changes (cf. e.g., Santiago and Bifano 2005; Huchzermeier and Loch 2001). In the presence of a later uncertainty reduction possibility, however, this strategy becomes suboptimal. It leads to costly improvements as well as to a precipitate abandonment in low performance states of the project, albeit these actions are not the optimal choice under consideration of the actual market performance requirements. In expectation of a later updating possibility, it is generally optimal to exercise less managerial options prior to the update. This is consistent with the options intuition that in the presence of later options, the earlier option may be exercised less likely and hence, is less valuable than in isolation. As the market uncertainty is reduced after the update and management knows thereafter more precisely which performance the customers will actually require at the time of the product launch, it is optimal to invest instead in appropriate counter measures after the information gain.

Principle 3: Intermediate information is of substantial value, especially for projects with high market uncertainty.

The analysis of our valuation framework has shown that the value of the information update increases with the amount of uncertainty that can be reduced with it. This insight is in line with decision theory and corresponds to common intuition. Nevertheless, it has clear practical implications. In principle, the uncertainty reduction is independent of the prior variance. In practice, however, the reduction of an already very low prior uncertainty by a certain percentage will significantly be harder to achieve compared to a high prior uncertainty level since a certain amount of uncertainty will always remain regardless of the available market studies and forecast methods (cf. e.g., Beardsley and Mansfield 1978). Thus, management should consider an information update particularly for development projects with a high prior market requirement uncertainty. The upper limit for the expenditures to acquire additional information during the development process thereby depends on the determined expected value of information. However, up to this amount, it is valuable to invest in additional information. In other words, investing more creates value. This result runs counter to financial option pricing theory intuition and contributes to better decision making in NPD projects with high uncertainty.

Principle 4: Invest in development flexibility at late stages to increase the possibility to respond to the updated market requirements.

The value of flexibility to respond to updated market information depends on the underlying development cost structure. The level of the improvement costs in the remaining development stages determines whether it is valuable to invest in corrective actions in case of a requirement increase. We have shown that it is valuable to reduce later improvement costs at the expense of higher upfront investments. A similar finding is derived by Loch and Terwiesch (2005). They propose to change the cost structure of the project by influencing the task architecture in order to improve the response to preliminary information. This insight is also supported by MacCormack and Verganti (2003) who empirically show that only projects with investments in design flexibility during the early development stages are likely to benefit from late design changes.

Such flexibility can be achieved through an initially broader development architecture or a more modular design (Krishnan and Bhattacharya 2002). Especially the latter is often regarded as an optimal strategy because it allows to independently adjust the components or subsystems to the requirement changes of the customers. This approach thus facilitates the cost reduction of design changes at late stages of the development process compared to traditional design architectures (Ulrich 1995; von Hippel 1990). But modularity also has its limits. Especially when new technologies are involved, very high degrees of design modularity in NPD projects generally reduce the system performance (Ethiraj and Levinthal 2004; Ulrich and Ellison 1999). Thus, a mix of appropriate measures should be selected in order to increase the development flexibility for the desired purpose.

Principle 5: Implement effective information generation mechanisms to obtain reliable market information early and inexpensively.

The structure of the updating costs also determines the ideal updating point in time. We have shown that for – in terms of time value of money – constant updating costs, the expected project value as well as the expected value of the information update decrease if the update of the market requirement distribution is postponed. The reason is that in case of such a postponement, management has less development stages left for taking appropriate counter measures to respond to the latest market requirement information. Thus, in this special case, it is optimal to update at the earliest feasible point in time. In all other cases, however, the ideal updating moment depends on the actual costs of obtaining additional information. It therefore decreases with a reduction of the information acquisition expenses. In other words, the lower the updating costs, the earlier the ideal updating point in time. Hence, besides having a flexible design and development process, management should focus on effective and cost efficient methods of information generation during the entire development process. Early customer integration through lead users, for example, or the evaluation of related products that have been recently launched in the market are effective means to obtain valuable information at the different development stages (cf. e.g., von Hippel 1986; Urban and von Hippel 1988).

6.3 Possible Extensions and Future Research

The presented model improves decision making under uncertainty in complex environments like the development of new products by combining two previously unlinked methodologies. This allows to determine the value as well as the implications of information updates on managerial decisions. Focusing on the setup of the model and the underlying assumptions that have been made in order to derive some general insights, the following extensions of the model appear to be promising for future research.

One possible extension of the decision model would be to consider multiple updating possibilities of the market requirement distribution during the development process. This extension is, from a Bayesian point of view, straightforward. The posterior distribution obtained after the first update becomes the prior distribution of the next updating period, which is then updated again with the information acquired during this stage, and so forth. The challenge is, however, on the decision making side. The information obtained during the first period determines the available information in the second one and hence, influences all related decisions. The determination of the ideal updating point in time therefore has to occur in a backward recursive manner relying on several assumptions regarding the potentially observable signals. If such assumptions are justifiable for the specific project setting, the model can provide valuable insights for management regarding the optimal timing of multiple information acquisitions and the optimal managerial response to the continuous information gain.

A possibility to further increase the practical applicability of the developed framework would be to relax some of the underlying assumptions. In order to reduce the complexity for the analysis of the model and to prove some insights in closed form, we assumed that the performance variability is limited to one performance state. For practical applications of the model, however, any degree of performance variability could be integrated. In the same manner, additional corrective actions, like improvements by more than one performance level, could be considered in order to further investigate the impact of development flexibility on the project value in the presence of an updating possibility. This extension can be conducted independently of the performance variability adjustment if improvements are limited to whole performance levels.

Although we grounded the base case of our numerical analysis on a real-life investment project, the data originated from an ongoing development project. Thus, an evaluation in retrospect was not possible. An ex post evaluation of the project at the different development stages with the at these points in time available information would be interesting in order to determine the full potential of our framework. Simultaneously, one could consider not only one performance parameter that determines the project payoff and hence, its value as in our case, but apply the developed multidimensional version of the model to a real-life project. As long as the different dimensions are independent of each other, all insights and managerial implications derived from the presented one-dimensional case remain their validity and can be applied.

Very interesting, but more distant directions for future research would be to study managerial flexibility for several NPD projects in parallel or to incorporate externalities of information acquisition, like market interactions. The latter has been addressed, for example, by Grenadier (1999) who studies option exercise strategies with imperfect information where firms with asymmetric private information update their beliefs by observing each other's investment decisions. Future research could incorporate such dynamics and limitations of the information acquisition and study their implications on the strategic value of information updates in NPD projects.

As product development gains in importance for most companies and the risk as well as the cost of NPD projects simultaneously increases, the need for such research will even grow. Hence, more quantitative models of the type developed here will be needed for further supporting managerial decision making and enhancing the management of uncertainty in NPD projects.

A

Statistical Distributions

In the subsequent section, we provide a summary of the key statistical distributions that are used throughout the thesis. All properties presented in Chapter 3 as well as the derived results presented in Chapter 4 and Chapter 5 are based on the density functions as defined below. We will limit this listing to the definition of the density function, the range of parameter values, and its most interesting moments. (cf. Carlin and Louis 2000, p. 323 ff.)

A.1 Univariate Distributions

A.1.1 Normal - $N(\mu, \sigma^2)$

A continuous random quantity X has a *normal distribution* with parameters μ and σ^2 ($\mu \in \Re, \sigma > 0$) if its density function is

$$f(x|\mu, \sigma^2) = \frac{1}{\sigma\sqrt{2\pi}} \exp\left[-\frac{(x-\mu)^2}{2\sigma^2}\right].$$ (A.1)

The mean and variance are $E(X) = \mu$ and $Var(X) = \sigma^2$, respectively.

A.1.2 Gamma - $G(\alpha, \beta)$

A continuous random quantity X has a *gamma distribution* with parameters α and β ($\alpha > 0, \beta > 0$) if its density function is

$$f(x|\alpha, \beta) = \frac{1}{\Gamma(\alpha)\beta^\alpha} x^{\alpha-1} e^{x/\beta}.$$ (A.2)

The mean and variance are $E(X) = \alpha\beta$ and $Var(X) = \alpha\beta^2$, respectively.

Note that $\Gamma(\cdot)$ denotes the *gamma function* defined by

$$\Gamma(\alpha) = \int_0^{+\infty} y^{\alpha-1} e^{-y} dy, \qquad \alpha > 0. \qquad (A.3)$$

A.1.3 Inverse Gamma - IG(α, β)

A continuous random quantity X has a *inverse gamma distribution* with parameters α and β ($\alpha > 0, \beta > 0$) if its density function is

$$f(x|\alpha, \beta) = \frac{1}{\Gamma(\alpha)\beta^\alpha x^{\alpha+1}} e^{-1/(\beta x)}. \qquad (A.4)$$

The mean and variance are

$$E(X) = \frac{1}{\beta(\alpha-1)}, \qquad \text{if } \alpha > 1$$

$$Var(X) = \frac{1}{\beta^2(\alpha-1)^2(\alpha-2)}, \qquad \text{if } \alpha > 2.$$

Note: $1/X \sim G(\alpha, \beta)$.

A.1.4 t (or Student's t) - St(μ, σ^2, α)

A continuous random quantity X has a *t distribution* with location parameter μ, scale parameter σ^2, and α degrees of freedom ($\mu \in \Re, \sigma > 0, \alpha > 0$) if its density function is

$$f(x|\mu, \sigma^2, \alpha) = \frac{\Gamma[(\alpha+1)/2]}{\sigma\sqrt{\alpha\pi}\Gamma(\alpha/2)} \left[1 + \frac{1}{\alpha}\frac{(x-\mu)^2}{\sigma^2}\right]^{-\frac{(\alpha+1)}{2}}. \qquad (A.5)$$

The mean and variance are

$$E(X) = \mu, \qquad \text{if } \alpha > 1$$

$$Var(X) = \frac{\alpha\sigma^2}{\alpha-2}, \qquad \text{if } \alpha > 2.$$

A.2 Multivariate Distributions

A.2.1 Multivariate Normal - $N_k(\mu, \Sigma)$

A k-dimensional, continuous random vector $X = (X_1, \ldots, X_k)'$ ($x \in \Re^k$) has a *multivariate normal distribution* of dimension k, with mean vector

$\mu = (\mu_1, \ldots, \mu_k)'$ and covariance matrix Σ, where $\mu_i \in \Re$ and Σ is a $k \times k$ symmetric, positive-definite matrix, if its density function is

$$f(x|\mu, \Sigma) = \frac{1}{(2\pi)^{p/2}(det\Sigma)^{1/2}} \exp\left[-\frac{1}{2}(x-\mu)'\Sigma^{-1}(x-\mu)\right]. \qquad (A.6)$$

The mean vector and covariance matrix are $E(X) = \mu$ and $Var(X) = \Sigma$, respectively, where $Var(X_i) = \sigma_{ii}$ and $Cov(X_i, X_j) = \sigma_{ij}$, so that $\Sigma = (\sigma_{ij})$.

A.2.2 Multivariate t - $St_k(\mu, \Sigma, \alpha)$

A k-dimensional, continuous random vector $X = (X_1, \ldots, X_k)'$ $(x \in \Re^k)$ has a *multivariate t distribution* of dimension k, with location vector $\mu = (\mu_1, \ldots, \mu_k)'$, scale matrix Σ, and α degrees of freedom, where $\mu_i \in \Re$, Σ is a $k \times k$ symmetric positive-definite matrix and $\alpha > 0$, if its density function is

$$f(x|\alpha, \mu, \Sigma) = \frac{\Gamma[(\alpha+k)/2]}{(det\Sigma)^{1/2}(\alpha\pi)^{k/2}\Gamma(\alpha/2)}\left[1 + \frac{1}{\alpha}(x-\mu)'\Sigma^{-1}(x-\mu)\right]^{-\frac{\alpha+k}{2}}.$$
$$(A.7)$$

The mean vector and covariance matrix are $E(X) = \mu$ (if $\alpha > 1$) and $Var(X) = (\alpha\Sigma)/(\alpha - 2)$ (if $\alpha > 2$), respectively.

A.2.3 Wishart - $Wi_k(\Omega, \alpha)$

Let $X_1, \ldots, X_n$ be a random sample of k-dimensional random vectors from a multivariate normal distribution with mean vector $\mu = 0$ and a $k \times k$ covariance matrix Σ. A $k \times k$ symmetric, positive-definite matrix V of random quantities which is defined by the equation

$$V = \sum_{i=1}^{n} X_i X_i' \qquad (A.8)$$

has a *Wishart distribution* of dimension k, with parameter matrix Ω and α degrees of freedom, where Ω is a $k \times k$, symmetric, nonsingular matrix and $\alpha > k - 1$, if the density function of the $k(k+1)/2$ dimensional random vectors of the distinct variables V_{ij} is

$$f(V|\alpha, \Omega) = c\frac{|V|^{(\alpha-k-1)/2}}{|\Omega|^{\alpha/2}} \exp\left[-\frac{1}{2}tr(\Omega^{-1}V)\right], \qquad (A.9)$$

where $tr(\cdot)$ denotes the trace of a matrix argument and the constant c has the following form:

$$c = \left[2^{(\alpha k)/2} \pi^{k(k-1)/4} \prod_{j=1}^{k} \Gamma\left(\frac{\alpha + 1 - j}{2} \right) \right]^{-1}. \tag{A.10}$$

The mean vector is $E(V) = \alpha \Omega^{-1}$. Further properties are $Var(V_{ij}) = \alpha \left(\Omega_{ij}^2 + \Omega_{ii}\Omega_{jj} \right)$ and $Cov(V_{ij}V_{kl}) = \alpha(\Omega_{ik}\Omega_{jl} + \Omega_{il}\Omega_{jk})$.

Note: The Wishart distribution is a multivariate generalization of the gamma distribution.

B

Supplement to Performance Variability Limitation

Our information updating valuation model is based on the decision model of Huchzermeier and Loch (2001). Contrary to their model, we limit the variability of the product performance to $N = 1$. In the subsequent section, we will explain this limitation and its implication for our model in greater detail and prove the validity of applying the properties derived by Huchzermeier and Loch to our case. For reasons of consistency, we will apply our notation also to their properties and proofs.

In their decision model, Huchzermeier and Loch generalize the performance variability to be spread over N states, i.e., the uncertainty of the development process at stage t, denoted by ω_t, is given as

$$\omega_t = \begin{cases} \frac{i}{2} & \text{with probability } \frac{p}{N} \\ -\frac{i}{2} & \text{with probability } \frac{1-p}{N} \end{cases} \quad \text{for } i = 1, \ldots, N. \quad \text{(B.1)}$$

They claim in Proposition 1 (p. 91), amongst other properties of the value function, that the project value $V_t(x)$ is convex-concave increasing in x if the payoff function $\Pi(x)$ is increasing in x as well. The proof of the corresponding lemma (Lemma 1, p. 99) is based on the argument that $\frac{1}{1+r}E[V_{t+1}(x + 1) - V_{t+1}(x)]$ increases and decreases in x since $V_{t+1}(x + 1) - V_{t+1}(x)$ does so due to the described convex-concavity of $V_{t+1}(x)$. In particular, they state that improvement is preferred over continuation in state x iff

$$\alpha_t < \frac{1}{1+r} E_{\omega_t} \left[V_{t+1}(x+1) - V_{t+1}(x) \right]$$

$$= \frac{1}{N(r+1)} \sum_{i=1}^{N} \left[p V_{t+1}\left(x+1+\frac{i}{2}\right) - (1-p)V_{t+1}\left(x+1-\frac{i}{2}\right) \right]$$

$$- \frac{1}{N(r+1)} \sum_{i=1}^{N} \left[p V_{t+1}\left(x+\frac{i}{2}\right) - (1-p)V_{t+1}\left(x-\frac{i}{2}\right) \right] \qquad \text{(B.2)}$$

and – by the convex-concavity of $V_{t+1}(x)$ – the right-hand side of Eq. B.2 first increases and then decreases in x.

However, Santiago and Vakili (2005) have shown that this has not necessarily to be the case. For certain cases, the difference function may have more than one maximum. The reason is that the average of several function each of which first increases and then decreases does not necessarily have the increasing-decreasing property. This discrepancy, however, can be resolved by limiting the product performance variability to $N = 1$. In this special case namely, the argument from above holds, i.e.,

$$\alpha_t < \frac{1}{1+r} E_{\omega_t} \left[V_{t+1}(x+1) - V_{t+1}(x) \right]$$

$$= \frac{1}{(r+1)} \left[p V_{t+1}\left(x+1+\frac{1}{2}\right) - (1-p)V_{t+1}\left(x+1-\frac{1}{2}\right) \right]$$

$$- \frac{1}{(r+1)} \left[p V_{t+1}\left(x+\frac{1}{2}\right) - (1-p)V_{t+1}\left(x-\frac{1}{2}\right) \right], \qquad \text{(B.3)}$$

since we now do not average over multiple functions anymore. Thus, the resulting averaged function has a single mode, i.e., is convex-concave increasing in x.

Although this assumption limits the uncertainty of the development process (modeled by the described binomial distribution) and thus, abstracts from the performance variability occurring in most real-life projects, it does not confine the practical applicability of the model too much. For the objective of our research, to combine real options valuation with Bayesian analysis and to study the resulting implications, the extent of the performance variability is circumstantial. In fact, the assumption allows us to reduce the complexity of our information updating valuation model noticeably and thus, enabling us to analyze the model with respect to the desired insights more generally. In addition and most of all, all other properties of the Huchzermeier and Loch model, in particular the property of the market performance variability, remain valid due to this limitation, which allow us to prove some of our findings in closed form and explain some numerical results more plausibly. However, for the sole purpose of an application to a real-life development project, the performance spread over more than $N = 1$ states can be easily integrated.

C

Supplement to Mean-Variance Update

In the following, we will illustrate the density functions of the simultaneous mean-variance updating example presented in Section 3.2.4.2. We will hereby present the prior as well as the posterior distribution of the marginal probability density functions of 1) the market requirement mean θ, 2) the market requirement variance σ^2, and 3) the market performance requirement d.

Table C.1. Parameters of prior distribution

Parameter	Value
α	2.8
β	0.28
μ	0
ν	1

Table C.1 specifies the parameters of the prior distributions based on the following given estimated moments of the unknown mean θ and unknown variance σ^2: $E(\theta) = 0$, $Var(\theta) = 2$, $E(\sigma^2) = 3$, and $Var(\sigma^2) = 2$.

The corresponding prior marginal density distributions the unknown mean θ, $\pi_1(\theta) = St\left(\theta|\mu, \frac{\nu}{\alpha\beta}, 2\alpha\right)$, and the unknown variance σ^2, $\pi_2(\sigma^2) = IG(\sigma^2|\alpha, \beta)$, are as plotted in Fig. C.1 and Fig. C.2. Fig. C.3 presents the corresponding prior marginal density distribution of the market performance requirement, $m(d) = St\left(d|\mu, \frac{\nu+1}{\alpha\beta}, 2\alpha\right)$.

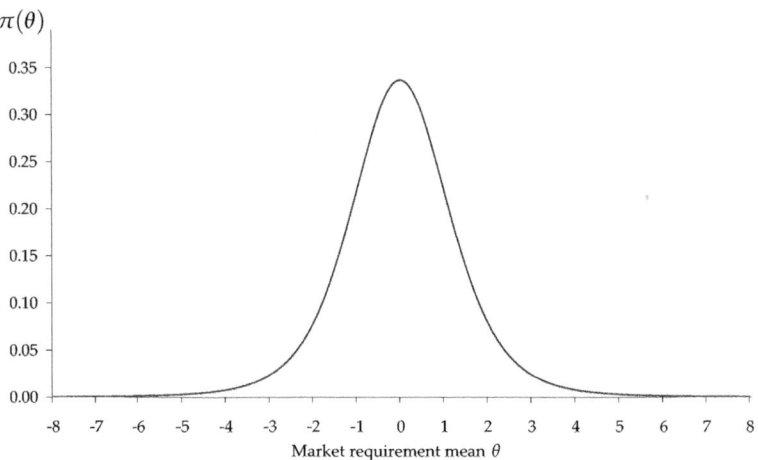

Fig. C.1. Prior marginal density distribution of mean

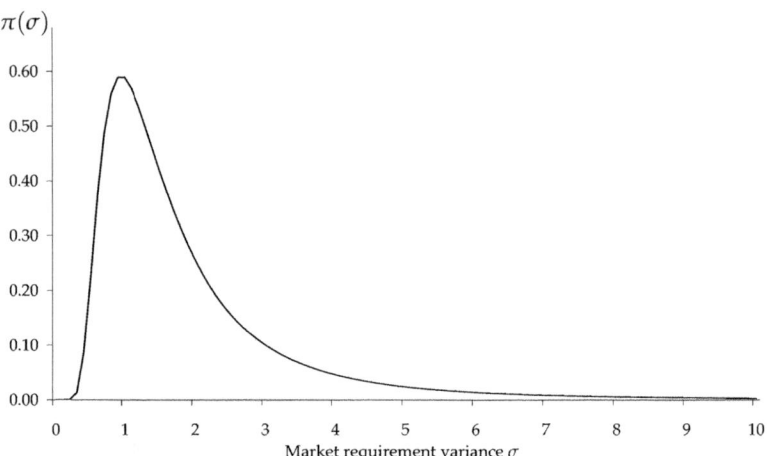

Fig. C.2. Prior marginal density distribution of variance

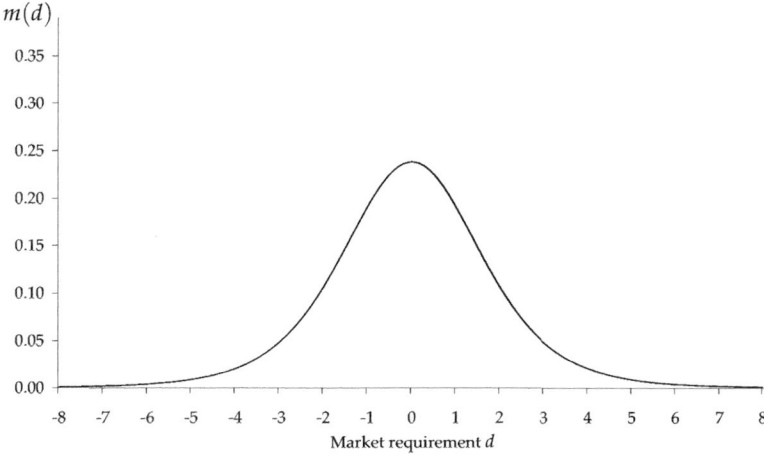

Fig. C.3. Prior marginal density distribution of market requirement

These initial estimates of the market requirement mean and the variance are updated based on the insights obtained from a follow-up study. As assumed in the example of Section 3.2.4.2, this study reveals a signal mean of $\bar{d} = 1.5$ with a variance of $\sum_{i=1}^{6}(d_i - \bar{d}) = 2$, i.e., $z = N(1.5; 2)$, based on a sample size of $n = 6$. With this information, we can determine the parameters of the posterior distribution as specified in Table C.2.

Table C.2. Parameters of posterior distribution for various sample sizes

	Value		
Parameter	$n = 6$	$n = 1$	$n = 10$
α'	11.6	3.3	7.8
β'	0.18	0.19	0.18
μ'	1.29	0.75	1.36

To illustrate the impact of the sample size n, we additionally determine the parameters of the posterior market requirement distribution for samples with a size of $n = 1$ and $n = 10$ (all other signal parameters remain constant). Table C.3 finally shows the corresponding moments of the posterior marginal density distributions for the mean and the variance.

Based on the above derived parameters of these posterior distributions, we can determine the marginal density function of the posterior

Table C.3. Moments of posterior distribution for various signals

Unknown parameter	Moment	$n = 6$	$n = 1$	$n = 10$
Mean θ	$E(\theta\|z)$	1.29	0.75	1.36
	$Var(\theta\|z)$	0.17	1.12	0.08
Variance σ^2	$E(\sigma^2\|z)$	1.16	2.25	0.83
	$Var(\sigma^2\|z)$	0.35	3.89	0.12

(The "Value" header spans the $n=6$, $n=1$, $n=10$ columns.)

mean, i.e., $\pi_1(\theta|z) = St\left(\theta|\mu', \left[(n + \frac{1}{\nu})(\alpha + \frac{n}{2})\beta'\right]^{-1}, 2\alpha + n\right)$, as well as
of the posterior variance, i.e., $\pi_2(\sigma^2|z) = IG\left(\sigma^2|\alpha + \frac{n}{2}, \beta'\right)$. Fig. C.4 and
Fig. C.5 plot the corresponding distributions for the signal with sample
size $n = 6$. The resulting posterior marginal density function of the mar-
ket performance requirement, depicted in Fig. C.6, is given as $m(d|z) =$
$St\left(d|\mu', \left[\frac{(\nu^{-1}+n)(\alpha+\frac{n}{2})\beta'}{\nu^{-1}+n+1}\right]^{-1}, 2\alpha + n\right)$. We can see that the larger the sample size n, the more weight is on the signal, and thus, the sample and the
posterior mean converge. In other words, with an increasing sample size
the posterior mean draws nearer to the mean of the signal. Note that the
posterior variance simultaneously decreases.

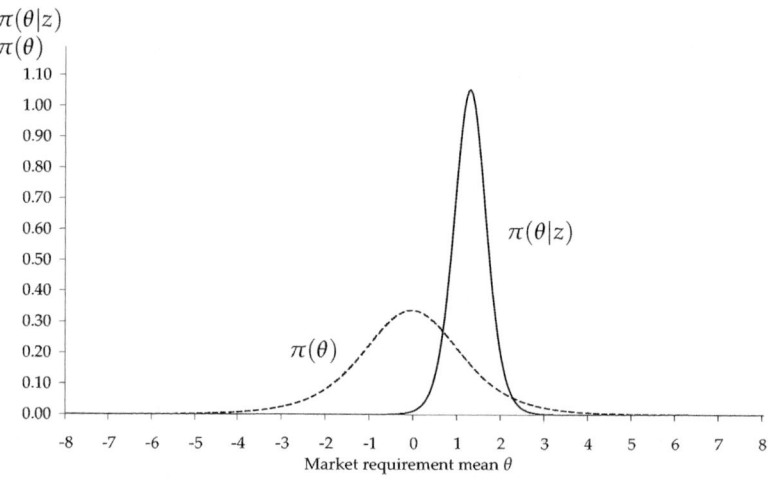

Fig. C.4. Posterior marginal density distribution of mean

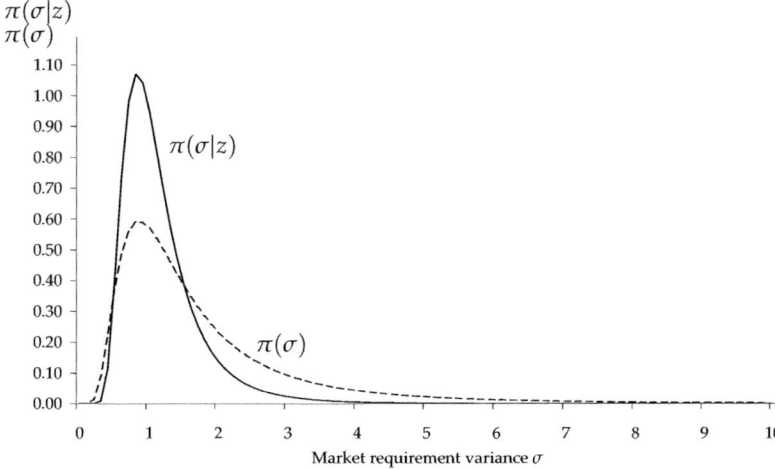

Fig. C.5. Posterior marginal density distribution of variance

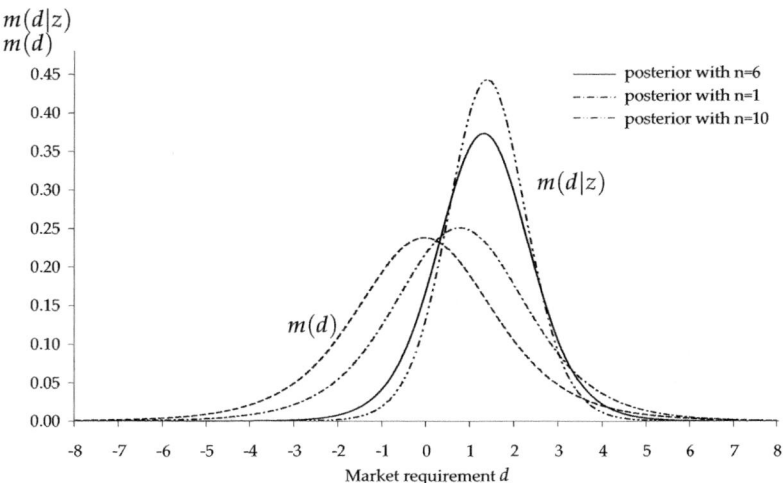

Fig. C.6. Posterior marginal density distribution of market requirement

D

Program Listing

In the following section, we provide the MATLAB code of the stochastic dynamic programs applied in Chapter 5 to run the numerical experiments for the comparative static analysis. We limit the listing to the definition of the variables and parameters as well as the algorithms for the determination of the different project values. Depending on the desired analysis, the subsequent algorithms allow to determine the project value, posterior project value, and the expected project value as well as the corresponding managerial policies and the (expected) value of information.[1]

D.1 Project Value

This algorithm allows to determine the project value in absence of any updating possibility as defined in Section 3.1.

---

[1]Note that contrary to our hitherto applied notation, starting point of the project is $t = 1$ since MATLAB cannot deal with variables starting with 0th component. The same holds true for the performance state x.

```
%%%%%%%%%%%%%%%%%%%%%% Definition of paramaters and variables %%%%%%%%%%%%%%%%%%%%%%%%%%%%%%%%%%%%%%%

T                       % Development periods
r                       % Discount rate

M                       % Maximum payoff
m                       % Minimum payoff
contcost=zeros(T);      % Continuation costs c_t
impcost=zeros(T);       % Improvement costs α_t

mu                      % Market requirement mean μ
st                      % Market requirement standard deviation σ

V=zeros(T+2,T+1);       % Project value V_t(x)
AV=zeros(T+2,T+1);      % Managerial policy a_t^x

%%%%%%%%%%%%%%%%%%%%%%%% Determination of project value %%%%%%%%%%%%%%%%%%%%%%%%%%%%%%%%%%%%%%%

for i=1:1:T+2
    V(i,T+1)=normcdf(((((T+2)/2)-(1*(i-1))),mu,(st))*(M-m); % for binomial tree with T=6
end % for i=1:1:T+2
```

```
for t=T+1:-1:2
    for i=1:t
        if i==1
            V(1,t-1)=max((V(1,t)+V(2,t))/(2*(1+r))-contcost(t-1),0);
        else  V(i,t-1)=max(max(((V(i-1,t)+V(i,t))/(2*(1+r))-contcost(t-1)-/
            impcost(t-1)),((V(i,t)+V(i+1,t))/(2*(1+r))-contcost(t-1))),0);
            if V(i,t-1)==((V(i-1,t)+V(i,t))/(2*(1+r))-contcost(t-1)-impcost(t-1))
                AV(i,t-1)=1;
            elseif V(i,t-1)==((V(i,t)+V(i+1,t))/(2*(1+r))-contcost(t-1))
                AV(i,t-1)=0;
            elseif V(i,t-1)==0
                AV(i,t-1)=-1;
            else AV(i,t-1)=9999;
            end %if V(i,t-1)==((V(i-1,t)+V(i,t))/(2*(1+r))-contcost(t-1)-impcost(t-1))
        end %if i==1
    end % for i=1:t
end %for t=T+1:-1:2
```

D.2 Posterior Project Value

This algorithm allows to determine the posterior project value based on an update of the market requirement mean. The prior project value is hereby determined with the algorithm presented in Appendix D.1.

```
%%%%%%%%%%%%%%%%%%%%%%%%%%%%% Definition of paramaters and variables %%%%%%%%%%%%%%%%%%%%%%%%%%%%%%%%%%%%

T                          % Development periods
r                          % Discount rate

M                          % Maximum payoff
m                          % Minimum payoff

contcost=zeros(T);         % Continuation costs $c_t$
impcost=zeros(T);          % Improvement costs $\alpha_t$

mu                         % Prior mean $\mu$
s                          % Spread of market requirements in target market $\sigma$
xi                         % Prior standard deviation of true market requirement mean $\zeta$
st                         % Prior market requirement standard deviation $\sigma_{prior}$
z                          % Signal
mup                        % Posterior mean $\mu'$
sp                         % Posterior market requirement standard deviation $\sigma'$

V=zeros(T+2,T+1);          % Prior project value $V_t^{prior}(x)$
AV=zeros(T+2,T+1);         % Prior managerial policy $a_{x,t}^{prior}$
EV=zeros(T+2,T+1);         % Posterior project value $V_t(x,z)$
EAV=zeros(T+2,T+1);        % Posterior managerial policy $a_t^x$
```

```
PV=zeros(T+2,T+1);             % Posterior project value based on prior managerial policy P_t(x,z)
IV=zeros(T+2,T+1);             % Value of information V_t^I(x,z)

%%%%%%%%%%%%%%%%%%% Determination of prior project value %%%%%%%%%%%%%%%%%%%%%%%%%%%%%

See algorithm of Appendix D.1

%%%%%%%%%%%%%%%% Determination of posterior market requirement distribution %%%%%%%%%%%%%%%%%%%%

st=st=sqrt(s^2+xi^2);
mup=(s^2*mu+xi^2*z)/(s^2+xi^2);
sp=sqrt(s^2+((s^2*xi^2)/(s^2+xi^2)));

%%%%%%%%%%%%%%%%% Determination of posterior project value %%%%%%%%%%%%%%%%%%%%%%%%%%%%

for i=1:1:T+2
    EV(i,T+1)=normcdf((((T+2)/2)-(1*(i-1))),mup,(sp))*(M-m); % for binomial tree with T=6
end % for i=1:1:T+2

for t=T+1:-1:2
    for i=1:t
```

```
if i==1
    EV(1,t-1)=(EV(1,t)+EV(2,t))/(2*(1+r))-contcost(t-1);
    PV(1,t-1)=(PV(1,t)+PV(2,t))/(2*(1+r))-contcost(t-1);
else  EV(i,t-1)=max(max(((EV(i-1,t)+EV(i,t))/(2*(1+r))-contcost(t-1)-↙
       impcost(t-1)),((EV(i,t)+EV(i+1,t))/(2*(1+r))-contcost(t-1))),0);
    if EV(i,t-1)==((EV(i-1,t)+EV(i,t))/(2*(1+r))-contcost(t-1)-impcost(t-1))
        EAV(i,t-1)=1;
    elseif EV(i,t-1)==((EV(i,t)+EV(i+1,t))/(2*(1+r))-contcost(t-1))
        EAV(i,t-1)=0;
    elseif EV(i,t-1)==0
        EAV(i,t-1)=-1;
    else EAV(i,t-1)=9999;
    end %if EV(i,t-1)==((EV(i-1,t)+EV(i,t))/(2*(1+r))
end %if i==1
if AV(i,t-1)==1
    PV(i,t-1)=((PV(i-1,t)+PV(i,t))/(2*(1+r))-contcost(t-1)-impcost(t-1));
    IV(i,t-1)=EV(i,t-1)-PV(i,t-1);
elseif AV(i,t-1)==0
    PV(i,t-1)=((PV(i,t)+PV(i+1,t))/(2*(1+r))-contcost(t-1));
    IV(i,t-1)=EV(i,t-1)-PV(i,t-1);
else PV(i,t-1)=0;
    IV(i,t-1)=EV(i,t-1)-PV(i,t-1);
end %if AV(i,t)==1
```

```
end % for i=1:t
end %for t=T+1:-1:2
```

%
%
%
%
%
%
%
%
%
%
%
%
%
%
%
%
%
%
%
%
%
%
%
%
%
%
%
%
%
%
%
%
%
%
%
%
%
%
%
%
%
%
%
%
%
%
%
%
%
%
%
%
%
%
%
%
%
%
%
%
%
%
%
%
%
%
%
%
%
%
%
%
%
%
%
%
%
%
%
%
%
%
%
%
%
%
%
%
%
%

D.3 Expected Project Value

This algorithm allows to determine the project value in expectation of a later update of the market requirement mean at different points in time, i.e., $\tau = 2, \ldots, T - 1$. The prior project value is hereby determined with the algorithm presented in Appendix D.1.

```
%%%%%%%%%%%%%%%%%%%%%%%%%% Definition of paramaters and variables %%%%%%%%%%%%%%%%%%%%%%%%%%%

T                          % Development periods
r                          % Discount rate

M                          % Maximum payoff
m                          % Minimum payoff
contcost = zeros(T);       % Continuation costs c_t
impcost = zeros(T);        % Improvement costs α_t
gamma=zeros(T);            % Updating costs

mu                         % Prior mean μ
s                          % Spread of market requirements in target market σ
xi                         % Prior standard deviation of true market requirement mean ξ
st                         % Prior market requirement standard deviation σ_prior
z                          % Signal
nz                         % Number of signal values considered for discretization
prob=zeros(T+1);           % Probability distribution for discretization
mup                        % Posterior mean μ'
sp                         % Posterior market requirement standard deviation σ'
ut                         % Updating point in time τ
nr                         % Number of program runs for different program settings
```

```
counter                          % Variable for different program runs

V=zeros(T+2,T+1);                % Prior project value V_t^{prior}(x)
AV=zeros(T+2,T+1);               % Prior managerial policy a_{x,t}^{prior}

EV=zeros(T+2,T+1,nz,T,nr);       % Posterior project value V_t(x,z)
EEV=zeros(T+2,T+1,nz,T,nr);      % Project values based for t=2,...,ut on expected optimal and /
                                   for t=ut+1,..T on optimal managerial policy (control variable)
AVP(T+2,T+1)                     % Posterior managerial policy a_t^x
PV=zeros(T+2,T+1,nz,T,nr);       % Project value based on prior managerial policy P_t(x,z)
IV=zeros(T+2,T+1,nz,T,nr);       % Value of information V_t^I(x,z)

IIV=zeros(T+2,T+1,nz,T,nr);      % (Expected) Value of information (control variable)
ERV=zeros(T+2,T+1,T,nr);         % Expected project value V_t^τ(x)
EPV=zeros(T+2,T+1,T,nr);         % Expected project value based on prior managerial/
                                   policy P_t^τ(x)
EIV=zeros(T+2,T+1,T,nr);         % Expected value of information V_t^{I,τ}(x)


%%%%%%%%%%%%%%%%%%%%%%%%%% Determination of prior project value %%%%%%%%%%%%%%%%%%%%%%%%%%%%%%

See algorithm of Appendix D.1
```

```
%%%%%%% Determination of expected project value for different updating points %%%%%%%

S=zeros(T+2,T+1);
SE=zeros(T+2,T+1);
v=zeros(T+2,T+1);
counter=counter+1;


for ut=2:1:T
prob = zeros(T+1);
prob(a)=1/nz;
a=0;

%%%%%%%%%%%%%%%%%%%%%%%% Determination of posterior project values %%%%%%%%%%%%%%%%%%%%%

for z=-4:1:4
  a=a+1;
  mup=(s^2*mu+xi^2*z)/(s^2+xi^2);
  sp=sqrt(s^2+((s^2*xi^2)/(s^2+xi^2)));

  for i=1:1:T+2
```

```
V(i,T+1)=normcdf((((T+2)/2)-(1*(i-1))),mup,sp)*(M-m);  % for binomial tree with T=6
EV(i,T+1,a,ut,counter)=V(i,T+1);
EEV(i,T+1,a,ut,counter)=V(i,T+1);
PV(i,T+1,a,ut,counter)=V(i,T+1);
end % for i=1:1:T+2

for t=T+1:-1:2
  for i=1:t
    if i==1
      v(1,t-1)=(v(1,t)+v(2,t))/(2*(1+r))-contcost(t-1);
      EV(1,t-1,a,ut,counter)=v(1,t-1);
      EEV(1,t-1,a,ut,counter)=v(1,t-1);
      PV(1,t-1,a,ut,counter)=(PV(1,t,a,ut,counter)+PV(2,t,a,ut,counter))/(2*(1+r))-/
                             contcost(t-1);
    else v(i,t-1)=max(max(((v(i-1,t)+v(i,t))/(2*(1+r))-contcost(t-1)-impcost(t-1)),/
                     ((v(i,t)+v(i+1,t))/(2*(1+r))-contcost(t-1))),0);
      EV(i,t-1,a,ut,counter)=v(i,t-1);
      EEV(i,t-1,a,ut,counter)=v(i,t-1);
      if AV(i,t-1)==1
        PV(i,t-1,a,ut,counter)=((PV(i-1,t,a,ut,counter)+PV(i,t,a,ut,counter))/
                               (2*(1+r))-contcost(t-1)-impcost(t-1));
        IV(i,t-1,a,ut,counter)=EV(i,t-1,a,ut,counter)-PV(i,t-1,a,ut,counter);
```

```
        elseif AV(i,t-1)==0
            PV(i,t-1,a,ut,counter)=((PV(i,t,a,ut,counter)+PV(i+1,t,a,ut,counter))/...
                                    (2*(1+r))-contcost(t-1));
            IV(i,t-1,a,ut,counter)=EV(i,t-1,a,ut,counter)-PV(i,t-1,a,ut,counter);
        else PV(i,t-1,a,ut,counter)=0;
            IV(i,t-1,a,ut,counter)=EV(i,t-1,a,ut,counter)-PV(i,t-1,a,ut,counter);
        end %if AV(i,t)==1

    end %if i==1
  end % for i=1:t
end %for t=T+1:-1:2

for k=1:ut+1
    S(k,ut)=prob(a)*v(k,ut)+S(k,ut);
    SE(k,ut)=prob(a)*PV(k,ut,a,ut,counter)+SE(k,ut);
end %for k=1:ut+1

v=zeros(T+2,T+1);
end %for z=-4:1:4

%%%%%%%%%%%%%%%%%%%%%%%% Determination of expected project value %%%%%%%%%%%%%%%%%%%%%%%%%%%
```

```
RV=zeros(T+2,T+1);
SRV=zeros(T+2,T+1);
test=zeros(T+2,T+1);

for i=1:T+2
    RV(i,ut)=S(i,ut)-gamma(ut);
    SRV(i,ut)=SE(i,ut);
    test(i,ut)=RV(i,ut);
    ERV(i,ut,ut,counter)=RV(i,ut);
    EPV(i,ut,ut,counter)=SE(i,ut);
end

for t=ut:-1:2
    for i=1:t
        if i==1
            RV(1,t-1)=(RV(1,t)+RV(2,t))/(2*(1+r))-contcost(t-1);
            ERV(1,t-1,ut,counter)=RV(1,t-1);
            EPV(1,t-1,ut,counter)=((EPV(1,t,ut,counter)+EPV(2,t,ut,counter))/(2*(1+r))-...
                contcost(t-1));
        else RV(i,t-1)=max(max(((RV(i-1,t)+RV(i,t))/(2*(1+r))-contcost(t-1)-impcost(t-1)),...
                ((RV(i,t)+RV(i+1,t))/(2*(1+r))-contcost(t-1))),0);
            ERV(i,t-1,ut,counter)=RV(i,t-1);
```

```
if RV(i,t-1)==((RV(i-1,t)+RV(i,t))/(2*(1+r))-contcost(t-1)-impcost(t-1))
    AVP(i,t-1)=1;
    for a=1:1:nz
        EEV(i,t-1,a,ut,counter)=((EEV(i-1,t,a,ut,counter)+↙
                          EEV(i,t,a,ut,counter))/(2*(1+r))-↙
                          contcost(t-1)-impcost(t-1));

        IIV(i,t-1,a,ut,counter)=EEV(i,t-1,a,ut,counter)-PV(i,t-1,a,ut,counter);
    end %a=1:1:nz


elseif RV(i,t-1)==((RV(i,t)+RV(i+1,t))/(2*(1+r))-contcost(t-1))
    AVP(i,t-1)=0;
    for a=1:1:nz
        EEV(i,t-1,a,ut,counter)=((EEV(i,t,a,ut,counter)+↙
                          EEV(i+1,t,a,ut,counter))/(2*(1+r))-↙
                          contcost(t-1)-impcost(t-1));↙

        IIV(i,t-1,a,ut,counter)=EEV(i,t-1,a,ut,counter)-↙
                          PV(i,t-1,a,ut,counter);

    end %for a=1:1:nz


elseif RV(i,t-1)==0
    AVP(i,t-1)=-1;
    for a=1:1:nz
        EEV(i,t-1,a,ut,counter)=0;
```

```
            IIV(i,t-1,a,ut,counter)=EEV(i,t-1,a,ut,counter)-↙
                                     PV(i,t-1,a,ut,counter);

        end %for a=1:1:nz

else AVP(i,t-1)=9999;
end %if V(i,t-1)==((V(i-1,t)+V(i,t))/(2*(1+r))-contcost(t-1)-impcost(t-1))

if AV(i,t-1)==1
    EPV(i,t-1,ut,counter)=((EPV(i-1,t,ut,counter)+EPV(i,t,ut,counter))/↙
                          (2*(1+r))-contcost(t-1)-impcost(t-1));
    EIV(i,t-1,ut,counter)=ERV(i,t-1,ut,counter)-EPV(i,t-1,ut,counter);

elseif AV(i,t-1)==0
    EPV(i,t-1,ut,counter)=((EPV(i,t,ut,counter)+EPV(i+1,t,ut,counter))/↙
                          (2*(1+r))-contcost(t-1));
    EIV(i,t-1,ut,counter)=ERV(i,t-1,ut,counter)-EPV(i,t-1,ut,counter);

else
    EPV(i,t-1,ut,counter)=0;
    EIV(i,t-1,ut,counter)=ERV(i,t-1,ut,counter)-EPV(i,t-1,ut,counter);
    end %if AV(i,t-1)==1

end %if i==1
```

```
end % for i=1:t
end %for t=ut:-1:2
```

%
%
%
%
%
%
%
%
%
%
%
%
%
%
%
%
%
%
%
%
%
%
%
%
%
%
%
%
%
%
%
%
%
%
%
%
%
%
%
%
%
%
%
%
%
%
%
%
%
%
%
%
%
%
%
%
%
%
%
%
%
%
%
%
%
%
%
%
%
%
%
%
%
%
%
%
%
%
%
%
%
%
%
%
%
%

References

ABRAHAM, B. and J. LEDOLTER (1983): *Statistical Methods for Forecasting*. Wiley, New York.

ADAMS, M. and D. BOIKE (2004): *PDMA Foundation's Comparative Performance Assessment Study (CPAS)*. Tech. rep., PDMA Foundation.

ALBACH, H. (1993): *Culture and Technical Innovation*. Walter de Gruyter, Berlin.

AMRAM, M. and N. KULATILAKA (1999): *Real Options - Managing Strategic Investment in an Uncertain World*. 1st ed., Harvard Business School Press, Boston, MA.

ANTELMAN, G. (1997): *Elementary Bayesian Statistics*. 1st ed., Edward Elgar Publishing Ltd., Cheltenham (UK).

ARMSTRONG, M., W. BAILEY, and B. COUET (2005): The Option Value of Acquiring Information in an Oilfield Production Enhancement Project. *Journal of Applied Corporate Finance* 17(2), pp. 99–104.

ATHEY, S., P. MILGROM, and J. ROBERTS (1998): *Robust Comparative Statics*. Monograph, Stanford University, Stanford.

ATUAHENE-GIMA, K. (1995): An explorative analysis of the impact of market orientation on new product performance. *Journal of product innovation management* 12(4), pp. 275–293.

AZOURY, K. S. (1985): Bayes Solution to Dynamic Inventory Models under Unknown Demand Distribution. *Management Science* 31(9), pp. 1150–1160.

AZOURY, K. S. and B. L. MILLER (1984): A Comparison of the Optimal Ordering Levels of Bayesian and Non-Bayesian Inventory Models. *Management Science* 30(8), pp. 993–1003.

BAECKER, P., U. HOMMEL, and H. LEHMANN (2003): Marktorientierte Investitionsrechnung bei Unsicherheit, Flexibilität, und Irreversibilität. In: U. HOMMEL, M. SCHOLICH, and P. BAECKER (eds.), *Reale Optionen*, Springer, Berlin, pp. 15–35.

BALBONTIN, A., B. YAZDANI, R. COOPER, and W. E. SOUDER (1999): New product development success factors in American and British firms. *International Journal of Technology Management* 17(3), pp. 259–280.

BALDWIN, C. Y. and K. B. CLARK (1998): *Modularity in Design: an Analysis based on the Theory of Real Options*. Working paper, Harvard Business School.

BALDWIN, C. Y. and K. B. CLARK (2002): The option value of Modularity in Design. In: C. Y. BALDWIN and K. B. CLARK (eds.), *Design Rules, Volume 1: The Power of Modularity*, MIT Press, Cambridge, MA.

BAYES, T. (1764): An Essay Toward Solving a Problem in the Doctrine of Chances. *Philosophical Transactions of the Royal Society of London* 53, pp. 370–418.

BEARDSLEY, G. and E. MANSFIELD (1978): A Note on the Accuracy of Industrial Forecasts of the Profitability of New Products and Processes. *Journal of Business* 51(1), pp. 127–135.

BELLALAH, M. (2001): Irreversibility, sunk costs and investment under incomplete information. *R&D Management* 31(2), pp. 115–126.

BERGER, J. O. (1985): *Statistical Decision Theory and Bayesian Analysis*. Springer Series in Statistics, 2nd ed., Springer-Verlag, New York.

BERNARDO, J. M. and A. F. M. SMITH (2000): *Bayesian Theory*. Wiley Series in Probability and Statistics, 1st ed., John Wiley & Sons, Chichester.

BERNOULLI, D. (1738): Specimen theoriae novae de mensura sortis. *Commentarii Academiae Scientiarum Imperialis Petropolitanae* 5, pp. 175–192, translated by L. Sommer (1954): Exposition of a new theory of the measurement of risk. *Econometrica* 22 (1), pp. 123-136.

BERTSEKAS, D. P. (2000): *Dynamic Programming and Optimal Control*. 2nd ed., Athena Scientific, Belmont, MA.

BHATTACHARYA, S., V. KRISHNAN, and V. MAHAJAN (1998): Managing New Product Definition in Highly Dynamic Environments. *Management Science* 44(11), pp. S50–S64.

BIRGE, J. R. (2000): Option Methods for Incorporating Risk into Linear Capacity Planning Models. *Manufacturing & Service Operations Management* 2(1), pp. 19–31.

BIRGE, J. R. and F. LOUVEAUX (1997): *Introduction to Stochastic Programming*. Springer, New York.

BLACK, F. and M. SCHOLES (1973): The Pricing of Options and Corporate Liabilities. *Journal of Political Economy* 81(3), pp. 637–654.

BONINI, C. P. (1977): Capital Investment under Uncertainty with Abandonment Options. *Journal of Financial and Quantitative Analysis* 12(1), pp. 39–51.

BREALEY, R. A. and S. C. MYERS (1996): *Principles of Corporate Finance*. 5th ed., McGraw-Hill, New York.

BRENNAN, M. J. and E. S. SCHWARTZ (1985): Evaluating Natural Resource Investments. *Journal of Business* 58(2), pp. 135–157.

BROCKHOFF, K. (1997a): *Forschung und Entwicklung*. 4th ed., Oldenbourg, München.

BROCKHOFF, K. (1997b): Wenn der Kunde stört – Differenzierungsnotwendigkeiten bei der Einbeziehung von Kunden in die Produktentwicklung. In: M. BRUHN and H. STEFFENHAGEN (eds.), *Marktorientierte Unternehmensführung*, Gabler, Wiesbaden, pp. 351–370.

BROCKHOFF, K. (1999): *Produktpolitik*. 4th ed., Lucius & Lucius, Stuttgart.

BROCKHOFF, K. (2000): Problems of Evaluating R&D Projects as Real Options. In: M. FRENKEL, U. HOMMEL, and M. RUDOLF (eds.), *Risk Management*, Springer, Berlin, pp. 203 – 212.

BRONSTEIN, I. N., K. A. SEMENDJAEV, G. MUSIOL, and H. MÜHLIG (1999): *Taschenbuch der Mathematik*. 4th ed., Harri Deutsch, Frankfurt am Main.

Brown, S. L. and K. M. Eisenhardt (1995): Product development: Past research, present findings, and future directions. *Academy of Management Review* 20(2), pp. 343–379.

Büyüközkan, G. and O. Feyzioglu (2004): A fuzzy-logic-based decision-making approach for new product development. *International Journal of Production Economics* 90(1), pp. 27–45.

Carbonell-Foulquié, P., J. L. Munuera-Alemán, and A. I. Rodríguez-Escudero (2004): Criteria employed for go/no-go decisions when developing successful highly innovative products. *Industrial Marketing Management* 33(4), pp. 307–316.

Carlin, B. P. and T. A. Louis (2000): *Bayes and Empirical Bayes Methods for Data Analysis.* Texts in Statistical Science, 2nd ed., Chapman & Hall, New York.

Chen, F. and J.-S. Song (2001): Optimal Policies for Multi-Echelon Inventory Problems with Markov Modulated Demand. *Operations Research* 49(2), pp. 226–234.

Childs, P. D., S. H. Ott, and A. J. Triantis (1998): Capital Budgeting for Interrelated Projects - A Real Options Approach. *Journal of Financial and Quantitative Analysis* 33(3), pp. 305–334.

Childs, P. D. and A. J. Triantis (1999): Dynamic R&D Investment Policies. *Management Science* 45(10), pp. 1359–1377.

Cooper, R. G. (1994): Third-Generation New Product Processes. *Journal of Product Innovation Management* 11(1), pp. 3–14.

Cooper, R. G., S. J. Edgett, and E. J. Kleinschmidt (2004): Benchmarking best NPD practices – II: Strategy, resource allocation and portfolio management. *Research Technology Management* 47(3), pp. 50–59.

Cooper, R. G. and E. J. Kleinschmidt (1994): Determinants of timeliness in product development. *Journal of Product Innovation Management* 11(5), pp. 381–391.

Cooper, R. G. and E. J. Kleinschmidt (1995): Benchmarking the Firm's Critical Success Factors in New Product Development. *Journal of Product Innovation Management* 12(5), pp. 374–391.

Cooper, R. G. and E. J. Kleinschmidt (1996): Winning businesses in product development: The critical success factors. *Research Technology Management* 39(4), pp. 18–29.

Copeland, T. E. and V. Antikarov (2001): *Real Options - A Practitioner's Guide.* 1st ed., Texere, New York.

Cortazar, G. and E. S. Schwartz (1998): Evaluating environmental investments: A real options approach. *Management Science* 44(8), pp. 1059–1070.

Cortazar, G., E. S. Schwartz, and J. Casassus (2001): Optimal exploration investment under price and geological-technical uncertainty: a real options model. *R&D Management* 31(2), pp. 181–189.

Cox, J. C. and S. A. Ross (1976): The Valuation of Options for Alternative Stochastic Processes. *Journal of Financial Economics* 3(1/2), pp. 145–166.

Cox, J. C., S. A. Ross, and M. Rubinstein (1979): Option Pricing: A Simplified Approach. *Journal of Financial Economics* 7(3), pp. 229–263.

Dahan, E. and H. Mendelson (2001): An Extreme-Value Model of Concept Testing. *Management Science* 47(1), pp. 102–116.

DAHAN, E. and V. SRINIVASAN (2000): The Predictive Power of Internet-Based Product Concept Testing Using Visual Depiction and Animation. *Journal of Product Innovation Management* 17(2), pp. 99–109.

DATAR, S. and C. JORDAN (1997): New product development structures and time-to-market. *Management Science* 43(4), pp. 452–465.

DEGROOT, M. H. (2004): *Optimal Statistical Decisions.* 1st ed., John Wiley & Sons, Chichester.

DIXIT, A. K. and R. S. PINDYCK (1994): *Investment under Uncertainty.* 1st ed., Princeton University Press, Princeton.

DVORETZKY, A., J. KEIFER, and J. WOLFOWITZ (1952): The inventory problem: II. Case of unknown distributions of demand. *Econometrica* 20(2), pp. 450–466.

EPPEN, G. D. and A. V. IYER (1997): Improved fashion buying with Bayesian updates. *Operations Research* 45(6), pp. 805–820.

ERNST, H. (2002): Success Factors of New Product Development: A Review of the Empirical Literature. *International Journal of Management Reviews* 4(1), pp. 1–40.

ETHIRAJ, S. K. and D. LEVINTHAL (2004): Modularity and Innovation in Complex Systems. *Management Science* 50(2), pp. 159–173.

FISHER, M. and A. RAMAN (1996): Reducing the Cost of Demand Uncertainty Through Accurate Response to Early Sales. *Operations Research* 44(1), pp. 87–99.

GARTNER, W. B. and R. J. THOMAS (1993): Factors Affecting New Product Forecasting Accuracy in New Firms. *Journal of Product Innovation Management* 10(1), pp. 35–52.

GAUR, V., A. GILONI, and S. SESHADRI (2005): Information Sharing in a Supply Chain Under ARMA Demand. *Management Science* 51(6), pp. 961–969.

GERWIN, D. and G. SUSMAN (1996): Special issue on concurrent engineering. *IEEE Transactions on Engineering Management* 43(2), pp. 118–123.

GESKE, R. S. (1979): The Valuation of Compound Options. *Journal of Financial Economics* 7(1), pp. 63–81.

GIRLICH, H.-J. and A. CHIKAN (2001): The origins of dynamic inventory modelling under uncertainty (the men, their work and connection with the Stanford Studies). *International Journal of Production Economics* 71(1-3), pp. 351–363.

GRAHAM, J. R. and C. R. HARVEY (2001): The theory and practice of corporate finance: evidence from the field. *Journal of Financial Economics* 60(2/3), pp. 187–243.

GRENADIER, S. R. (1999): Information revelation through option exercise. *Review of Financial Studies* 12(1), pp. 95–129.

GRIFFIN, A. (1997a): *Drivers of NPD Success: The PDMA Report 2007.* Tech. rep., PDMA Foundation.

GRIFFIN, A. (1997b): PDMA Research on New Product Development Practices: Updating Trends and Benchmarking Best Practices. *Journal of Product Innovation Management* 14(6), pp. 429–458.

GRIFFIN, A. and J. R. HAUSER (1993): The Voice of the Customer. *Marketing Science* 12(1), pp. 1–27.

GRIFFIN, A. and J. R. HAUSER (1996): Integrating R&D and Marketing: A Review and Analysis of the Literature. *Journal of Product Innovation Management* 13(3), pp. 191–215.

GRUNER, K. E. and C. HOMBURG (2000): Does Customer Interaction Enhance New Product Success? *Journal of Business Research* 49(1), pp. 1–14.

GURNANI, H. and C. S. TANG (1999): Note: Optimal Ordering Decisions with Uncertain Cost and Demand Forecast Updating. *Management Science* 45(10), pp. 1456–1462.

HA, A. Y. and E. L. PORTEUS (1995): Optimal timing of reviews in concurrent design for manufacturability. *Management Science* 41(9), pp. 1431–1447.

HALEY, G. T. and S. M. GOLDBERG (1995): Net Present Value Techniques and Their Effects on New Product Research. *Industrial Marketing Management* 24(3), pp. 177–190.

HAMMOND, J. H. (1990): *Quick Response in the Apparel Industry*. Harvard Business School Case N9-690-038, Cambridge, MA.

HAUSMAN, W. H. and R. PETERSON (1972): Multiproduct production scheduling for style goods with limited capacity, forecast revisions and terminal delivery. *Management Science* 18(7), pp. 370–383.

HE, H. and R. S. PINDYCK (2002): Investments in flexible production capacity. *Journal of Economic Dynamics and Control* 16(3-4), pp. 575–599.

HERATH, H. S. B. and C. S. PARK (2001): Real Options Valuation and Its Relationship to Bayesian Decision-Making Methods. *Engineering Economist* 46(1), pp. 1–32.

VON HIPPEL, E. (1986): Lead Users: A Source of Novel Product Concepts. *Management Science* 32(7), pp. 791–805.

VON HIPPEL, E. (1990): Task Partitioning: An Innovation Process Variable. *Research Policy* 19(5), pp. 407–418.

VON HIPPEL, E. (1992): *Adapting market research to the rapid evolution of needs for new products and services*. Working paper, Massachusetts Institute of Technology, Cambridge, MA.

HOLMAN, R., H.-W. KAAS, and D. KEELING (2003): The future of product development. *The McKinsey Quarterly* (3), pp. 28–40.

HUANG, G. Q. and K. L. MAK (1998): Re-Engineering the Product Development Process with 'design for X'. *Proceedings of the Institution of Mechanical Engineers – Part B – Engineering Manufacture* 212(4), pp. 259–268.

HUCHZERMEIER, A. and M. A. COHEN (1996): Valuing operational flexibility under exchange rate risk. *Operations Research* 44(1), pp. 100–114.

HUCHZERMEIER, A. and C. H. LOCH (2001): Project Management Under Risk: Using the Real Options Approach to Evaluate Flexibility in R&D. *Management Science* 47(1), pp. 85–101.

HULL, J. C. (2003): *Options, Futures and Other Derivatives*. 5th ed., Prentice Hall, Upper Saddle River, NJ.

HUNTER, A. (1990): *Quick Response in Apparel Manufacturing*. The Textile Institute, Manchester, U.K.

IANSITI, M. (1995): Technology Development and Integration: An Empirical Study of the Interaction between Applied Science and Product Development. *IEEE Transactions on Engineering Management* 42(3), pp. 259–269.

IGLEHART, D. L. (1964): The Dynamic Inventory Problem with Unknown Demand Distribution. *Management Science* 10(3), pp. 429–440.

INGERSOLL, J. E. and S. A. ROSS (1992): Waiting to invest: Investment and uncertainty. *Journal of Business* 65(1), pp. 1–29.

IYER, A. V. and M. E. BERGEN (1997): Quick Response in Manufacturer-Retailer Channels. *Management Science* 43(4), pp. 559–570.

JOHNSON, G. D. and H. E. THOMPSON (1975): Optimality of myopic inventory policies for certain dependent demand processes. *Management Science* 21(11), pp. 1303–1307.

KAHN, J. A. (1987): Inventories and the Volatility of Production. *American Economic Review* 77(4), pp. 667–679.

KAHN, K. B. (2002): An exploratory Investigation of new product forecasting practices. *Journal of Product Innovation Management* 19(2), pp. 133–143.

KALYANARAM, G. and V. KRISHNAN (1997): Deliberate product definition: Customizing the product definition process. *Journal of Marketing Research* 34(2), pp. 276–285.

KAMRAD, B. and R. ERNST (2001): An economic model for evaluating mining and manufacturing ventures with output yield uncertainty. *Operations Research* 49(5), pp. 690–699.

KARR, A. F. (1991): *Point Processes and Their Statistical Inference.* 2nd ed., Marcel Dekker, New York.

KESTER, W. C. (1984): Today's options for tomorrow's growth. *Harvard Business Review* 62(2), pp. 153–160.

KIM, H.-S. (2003): A Bayesian Analysis on the Effect of Multiple Supply Options in a Quick Response Environment. *Naval Research Logistics* 50, pp. 937–952.

KOGUT, B. and N. KULATILAKA (1994): Operating Flexibility, Global Manufacturing, and the Option Value of a Multinational Network. *Management Science* 40(1), pp. 123–139.

KRISHNAN, V. and S. BHATTACHARYA (2002): Technology Selection and Commitment in New Product Development: The Role of Uncertainty and Design Flexibility. *Management Science* 48(3), pp. 313–328.

KRISHNAN, V., S. D. EPPINGER, and D. E. WHITNEY (1997): A model-based framework to overlap product development activities. *Management Science* 43(4), pp. 437–451.

KRISHNAN, V. and C. H. LOCH (2005): Introduction to the Special Issue: Management of Product Innovation. *Production and Operations Management* 14(3), pp. 269–271.

KRISHNAN, V. and K. T. ULRICH (2001): Product Development Decisions: A Review of the Literature. *Management Science* 47(1), pp. 1–21.

KULATILAKA, N. (1988): Valuing the Flexibility of Flexible Manufacturing Systems. *IEEE Transactions on Engineering Management* 35(4), pp. 250–257.

KULATILAKA, N. (1993): The value of flexibility: The case of a dual-fuel industrial steam boiler. *Financial Management* 22(3), pp. 271–281.

KULATILAKA, N. (1995): Operating Flexibilities in Capital Budgeting: Substitutability and Complementarity in Real Options. In: L. TRIGEORGIS (ed.), *Real Options in Capital Investment*, 1st ed., Praeger, Westport, Connecticut, pp. 121–132.

KULATILAKA, N. and A. MARCUS (1992): Project Valuation under Uncertainty: When does DCF Fail? *Journal of Applied Corporate Finance* 5(3), pp. 92–100.

LANDER, D. M. and G. E. PINCHES (1998): Challenges to the Practical Implementation of Modeling and Valuing Real Options. *Quarterly Review of Economics & Finance* 38(4), pp. 537–567.

LARIVIERE, M. A. and E. L. PORTEUS (1999): Stalking Information: Bayesian Inventory Management with Unobserved Lost Sales. *Management Science* 45(3), pp. 346–363.

LAUX, H. (1991): *Entscheidungstheorie I.* 2nd ed., Springer, Berlin.

LEE, H. L., V. PADMANABHAN, and S. WHANG (1997): Information distortion in a supply chain: The bullwhip effect. *Management Science* 43(4), pp. 546–558.

LEE, H. L., K. C. SO, and C. S. TANG (2000): The Value of Information Sharing in a Two-Level Supply Chain. *Management Science* 46(5), pp. 626–643.

LEVITT, R. E., J. THOMSEN, T. R. CHRISTIANSEN, J. C. KUNZ, Y. JIN, and C. NASS (1999): Simulating Project Work Processes and Organizations: Toward a Micro-Contingency Theory of Organizational design. *Management Science* 45(11), pp. 1479–1495.

LINT, O. and E. PENNINGS (2001): An option approach to the new product development process: a case study at Philips Electronics. *R&D Management* 31(2), pp. 163–172.

LIPPMAN, S. A. and K. F. MCCARDLE (1987): Does Cheaper, Faster, or Better Imply Sooner in the Timing of Innovation Decision? *Management Science* 33(8), pp. 1058–1064.

LIPPMAN, S. A. and K. F. MCCARDLE (1991): Uncertain Search: A Model of Search among Technologies of Uncertain Values. *Management Science* 37(11), pp. 1474–1490.

LITTLE, A. D. (2004): *Innovation Excellence Studie.* Tech. rep., Arthur D. Little.

LOCH, C. H. and K. BODE-GREUEL (2001): Evaluating Growth Options as Sources of Value for Pharmaceutical Research Projects. *R&D Management* 31(2), pp. 231–248.

LOCH, C. H. and C. TERWIESCH (1998): Communication and uncertainty in concurrent engineering. *Management Science* 44(8), pp. 1032–1048.

LOCH, C. H. and C. TERWIESCH (2005): Rush and Be Wrong or Wait and Be Late? A Model of Information in Collaborative Processes. *Production and Operations Management* 14(3), pp. 331–343.

LOVEJOY, W. S. (1990): Myopic Policies for Some Inventory Models with Uncertain Demand Distributions. *Management Science* 36(6), pp. 724–738.

LYNN, G. S., S. P. SCHNAARS, and R. B. SKOV (1999): Survey of New Product Forecasting Practices in Industrial High Technology and Low Technology Businesses. *Industrial Marketing Management* 28(6), pp. 565–571.

MACCORMACK, A. and R. VERGANTI (2003): Managing the Sources of Uncertainty: Matching Process and Context in Software Development. *Journal of Product Innovation Management* 20(3), pp. 217–232.

MAHAJAN, V. and E. MULLER (1990): New Product Diffusion Models in Marketing: A Review and Directions for Research. *Journal of Marketing* 54(1), pp. 1–26.

MAHAJAN, V. and J. WIND (1992): New Product Models: Practice, Shortcomings and Desired Improvements. *Journal of Product Innovation Management* 9(2), pp. 128–139.

MAMER, J. W. and K. F. MCCARDLE (1987): Uncertainty, Competition, and the Adoption of New Technology. *Management Science* 33(2), pp. 161–177.

MARSCHAK, J. and R. RADNER (1972): *Economic Theory of Teams*. 1st ed., Yale University Press, New Haven.

MARSCHAK, T. and R. R. NELSON (1962): Flexibility, Uncertainty, and Economic Theory. *Metroeconomica* 14, pp. 42–58.

MARTZOUKOS, S. H. and L. TRIGEORGIS (2001): *Resolving a Real Options Paradox with Incomplete Information: After All, Why Learn?*. Working paper, University of Cyprus, Nicosia Cyprus.

MASON, S. and R. C. MERTON (1985): The Role of Contingent Claims Analysis in Corporate Finance. In: E. ALTMAN and M. SUBRAHMANYAM (eds.), *Recent Advances in Corporate Finance*, Irwin, Homewood, IL, pp. 7–54.

McCARDLE, K. F. (1985): Information Acquisition and the Adoption of New Technology. *Management Science* 31(11), pp. 1372–1389.

McDONALD, R. L. (2003): *Derivatives Markets*. Addison Wesley, Bosten, MA.

McDONALD, R. L. and D. R. SIEGEL (1985): Investment and the Valuation of Firms When There Is an Option to Shut Down. *International Economic Review* 26(2), pp. 331–349.

McDONALD, R. L. and D. R. SIEGEL (1986): The Value of Waiting to Invest. *Quarterly Journal of Economics* 101(4), pp. 707–727.

MERKHOFER, M. W. (1977): The Value of Information Given Decision Flexibility. *Management Science* 23(7), pp. 716–727.

MERTON, R. C. (1973): Theory of rational option pricing. *Bell Journal of Economics & Management Science* 4(1), pp. 141–183.

MEYER, M. H. and J. M. UTTERBACK (1995): Product development cycle time and commercial success. *IEEE Transactions on Engineering Management* 42(4), pp. 297–304.

MILGROM, P. and C. SHANNON (1994): Monotone Comparative Statics. *Econometrica* 62(1), pp. 157–180.

MILLER, L. T. and C. S. PARK (2002): Decision Making under Uncertainty - Real Options to the Rescue? *Engineering Economist* 47(2), pp. 105–150.

MILLER, L. T. and C. S. PARK (2005): A Learning Real Options Framework with Application to Process Design and Capacity Planning. *Production and Operations Management* 14(1), pp. 5–20.

MISHRA, S., D. KIM, and D. H. LEE (1996): Factors Affecting New Product Success: Cross-Country Comparisons. *Journal of Product Innovation Management* 13(6), pp. 530–550.

MONTOYA-WEISS, M. M. and R. CALANTONE (1994): Determinants of New Product Performance: A Review and Meta-Analysis. *Journal of Product Innovation Management* 11(5), pp. 397–417.

MURMANN, P. A. (1994): Expected Development Time Reductions in the German Mechanical Engineering Industry. *Journal of Product Innovation Management* 11(3), pp. 236–252.

MURTO, P. (2004): Valuing Options to Learn: Optimal Timing of Information Acquisition. In: *Proceedings of the 8th Annual International Conference on Real Options*, Montréal, Canada.

MYERS, S. C. (1976): Using simulation for risk analysis. In: S. C. MYERS (ed.), *Modern Developments in Financial Management*, Praeger, New York.

MYERS, S. C. (1977): Determinants of Corporate Borrowing. *Journal of Financial Economics* 5(2), pp. 147–176.

MYERS, S. C. (1984): Finance Theory and Financial Strategy. *Interfaces* 14(1), pp. 126–137.

MYERS, S. C. and S. MAJD (1990): Abandonment value and project life. *Advances in Futures and Options Research* 4(1-21).

NAHMIAS, S. (1997): *Production and Operations Analysis.* 3rd ed., Irwin, Homewood, IL.

VON NEUMANN, J. and O. MORGENSTERN (1944): *Theory of Games and Economic Behavior.* Princeton University Press, Princeton, NJ.

NEWTON, D. P., D. A. PAXSON, and A. W. PEARSON (1996): Real R&D options. In: A. BELCHER, J. HASSARD, and S. J. PROCTER (eds.), *R&D Decisions: Strategy, Policy and Innovations,* Routledge, London, pp. 273–282.

NUSSBAUM, B., R. BERNER, and D. BRADY (2005): Get Creative. *Business Week* (3945), pp. 60–68.

OTTUM, B. D. and W. L. MOORE (1997): The role of market information in new product success/failure. *Journal of Product Innovation Management* 14(4), pp. 258–273.

PICH, M. T., C. H. LOCH, and A. DE MEYER (2002): On Uncertainty, Ambiguity, and Complexity in Project Management. *Management Science* 48(8), pp. 1008–1023.

PINDYCK, R. S. (1988): Irreversible Investment, Capacity Choice, and the Value of the Firm. *American Economic Review* 78(5), pp. 969–985.

POLK, R., R. E. PLANK, and D. A. REID (1996): Technical Risk and New Product Success: An Empirical Test in High Technology Business Markets. *Industrial Marketing Management* 25(6), pp. 531–543.

RIEK, R. F. (2001): From experience: Capturing hard-won NPD lessons in checklists. *Journal of Product Innovation Management* 18(5), pp. 301–313.

ROSSI, P. E., G. M. ALLENBY, and R. MCCULLOCH (2005): *Bayesian Statistics and Marketing.* 1st ed., John Wiley & Sons, Hoboken, NJ.

ROTHWELL, R., C. FREEMAN, and A. HORLSEY (1974): SAPPHO updated - project SAPPHO phase II. *Research Policy* 3(3), pp. 258–291.

SANTIAGO, L. P. and T. G. BIFANO (2005): Management of R&D Projects Under Uncertainty: A Multidimensional Approach to Managerial Flexibility. *IEEE Transactions on Engineering Management* 52(2), pp. 269–280.

SANTIAGO, L. P. and P. VAKILI (2005): On the Value of Flexibility in R&D Projects. *Management Science* 51(8), pp. 1206–1218.

SAVAGE, L. J. (1954): *The Foundations of Statistics.* John Wiley & Sons, New York.

SCARF, H. E. (1959): Bayes Solutions of the Statistical Inventory Problem. *Annals of Mathematical Statistics* 30(2), pp. 490–508.

SCARF, H. E. (1960): Some Remarks on Bayes Solutions of the Statistical Inventory Problem. *Naval Research Logistics Quarterly* 7, pp. 591–596.

SCHRÖDER, H.-H. and A. J. JETTER (2003): Integrating market and technological knowledge in the fuzzy front end: an FCM-based actions support system. *International Journal of Technology Management* 26(5/6), pp. 517–539.

SESHADRI, S. and M. SUBRAHMANYAM (2005): Introduction to the Special Issue on "Risk Management in Operations". *Production and Operations Management* 14(1), pp. 1–4.

SETHI, S. P., H. YAN, and H. ZHANG (2005): *Inventory and supply chain management with forecast updates*. Springer, New York.

SHELLEY, C. J. and D. R. WHEELER (1991): New product forecasting horizons and accuracy. *Review of Business* 12(4), pp. 13–18.

SMITH, J. E. and K. F. McCARDLE (1998): Valuing oil properties: Integrating option pricing and decision analysis approaches. *Operations Research* 46(2), pp. 198–217.

SMITH, J. E. and R. F. NAU (1995): Valuing risky projects: Option pricing theory and decision analysis. *Management Science* 41(5), pp. 795–816.

SMITH, R. P. and S. D. EPPINGER (1997): A predictive model of sequential iteration in engineering design. *Management Science* 43(8), pp. 1104–1120.

SOBEK, D. K., A. C. WARD, and J. K. LIKER (1999): Toyota's principles of set-based concurrent engineering. *Sloan Management Review* 40(2), pp. 67–83.

SOMMER, S. C. and C. H. LOCH (2004): Selectionism and Learning in Projects with Complexity and Unforeseeable Uncertainty. *Management Science* 50(10), pp. 1334–1347.

SOUDER, W. E. and R. K. MOENAERT (1992): Integrating Marketing and R&D Project Personnel within Innovation Projects: An Information Uncertainty Model. *Journal of Management Studies* 29(4), pp. 485–512.

SRINIVASAN, V., W. S. LOVEJOY, and D. BEACH (1997): Integrated product design for marketability and manufacturing. *Journal of Marketing Research* 34(1), pp. 154–163.

TAKEUCHI, H. and I. NONAKA (1986): The new new product development game. *Harvard Business Review* 64(1), pp. 137–146.

TERWIESCH, C., C. H. LOCH, and A. DE MEYER (2002): Exchanging Preliminary Information in Concurrent Engineering: Alternative Coordination Strategies. *Organization Science* 13(4), pp. 402–420.

THOMKE, S. H. (1997): The role of flexibility in the development of new products: An empirical study. *Research Policy* 26(1), pp. 105–119.

THOMKE, S. H. (1998): Managing experimentation in the design of new products. *Management Science* 44(6), pp. 743–763.

THOMKE, S. H. and D. E. BELL (2001): Sequential Testing in Product Development. *Management Science* 47(2), pp. 308–323.

TRIGEORGIS, L. (1993): The Nature of Option Interactions and the Valuation of Investments with Multiple Real Options. *Journal of Financial & Quantitative Analysis* 28(1), pp. 1–20.

TRIGEORGIS, L. (1996): *Real Options - Managerial Flexibility and Strategy in Resource Allocation*. 4th ed., The MIT Press, Cambridge, MA.

TRIGEORGIS, L. and S. P. MASON (1987): Valuing Managerial Flexibility. *Midland Corporate Finance Journal* 5(1), pp. 14–21.

TULL, D. S. (1967): The Relationship of Actual and Predicted Sales and Profits in New-Product Introductions. *Journal of Business* 40(3), pp. 233–250.

TULL, D. S. and H. C. RUTEMILLER (1968): A Note on the Relationship of Actual and Predicted Sales and Profits in New-Product Introductions. *Journal of Business* 41(3), pp. 385–387.

ULRICH, K. T. (1995): The role of product architecture in the manufacturing firm. *Research Policy* 24(3), pp. 419–440.

ULRICH, K. T. (2001): Introduction to the Special Issue on Design and Development. *Management Science* 47(1), pp. v–vi.

ULRICH, K. T. and D. J. ELLISON (1999): Holistic Customer Requirements and the Design-Select Decision. *Management Science* 45(5), pp. 641–658.

ULRICH, K. T. and S. D. EPPINGER (2004): *Product Design and Development*. 3rd ed., McGraw-Hill/Irwin, New York.

UPTON, D. M. (1995): Flexibility as process mobility: The Management of plant capabilities for quick response manufacturing. *Journal of Operations Management* 12(3/4), pp. 205–224.

URBAN, G. I. and E. VON HIPPEL (1988): Lead User Analyses for the Development of New Industrial Products. *Management Science* 34(5), pp. 569–582.

VAN MIEGHEM, J. A. (1998): Investment strategies for flexible resources. *Management Science* 44(8), pp. 1071–1078.

VEINOTT, A. F. (1965): Optimal Policy for a Multi-Product, Dynamic, Nonstationary Inventory Problem. *Management Science* 12(3), pp. 206–222.

WALDMANN, K.-H. (1979): Numerical aspects in Bayesian inventory control. *Zeitschrift für Operations Research* 23, pp. 49–60.

WALL, M. B., K. T. ULRICH, and W. C. FLOWERS (1992): Evaluating prototyping technologies for product design. *Research in Engineering Design* 3(3), pp. 163–177.

WIND, J. and V. MAHAJAN (1997): Issues and Opportunities in New Product Development: An Introduction to the Special Issue. *Journal of Marketing Research* 34(1), pp. 1–12.

WITT, P. (2003): Die Bedeutung des Realoptionsansatzes für Gründungsunternehmen. In: U. HOMMEL, M. SCHOLICH, and P. BAECKER (eds.), *Reale Optionen*, Springer, Berlin, pp. 121–141.

ZAHAY, D., A. GRIFFIN, and E. FREDERICKS (2004): Sources, uses, and forms of data in the new product development process. *Industrial Marketing Management* 33(7), pp. 658–666.

ZANGWILL, W. I. (1992): Concurrent Engineering: Concepts and Implementation. *IEEE Engineering Management Review* 20(4), pp. 40–52.

Index

GPSR Compliance

*The European Union's (EU) General Product Safety Regulation (GPSR)
is a set of rules that requires consumer products to be safe and our
obligations to ensure this.*

*If you have any concerns about our products, you can contact us on
ProductSafety@springernature.com*

In case Publisher is established outside the EU, the EU authorized
representative is:

Springer Nature Customer Service Center GmbH
Europaplatz 3
69115 Heidelberg, Germany

Batch number: 09490872

Printed by Printforce, the Netherlands